Smart Card Security
and Applications

For a complete listing of the *Artech House Telecommunications Library*, turn to the back of this book.

Smart Card Security and Applications

Mike Hendry

Artech House
Boston • London

Library of Congress Cataloging-in-Publication Data
Hendry, Mike.
    Smart card security and applications / Mike Hendry.
        p.   cm.
    Includes bibliographical references (p.   ) and index.
    ISBN 0-89006-953-0 (alk. paper)
    1. Smart cards—Security measures.   I. Title.
TK7895.S62N46   1997
006—dc21   97-22713
                                                                            CIP

British Library Cataloguing in Publication Data
Hendry, Mike
    Smart card security and applications
    1. Smart cards   2. Smart cards—Industrial applications   3. Security systems
    I. Title.
    621.3'8928
    ISBN 0-89006-953-0

Cover design by Darrell Judd
**Cover images courtesy of Amphenol-Tuchel Electronics GmbH and
    Landis & Gyr Communications Corp.**

International Standard Book Number: 0-89006-953-0
Library of Congress Catalog Card Number: 97-22713

10 9 8 7 6 5 4 3 2 1

Contents

Foreword

THE LAST FEW YEARS have seen the most dynamic changes in payment card technology since the first magnetic stripe cards were introduced in the late 1960s. Microchip technology has transformed the way in which banks and other financial institutions look to offer their services. It provides cardholders with the ability to store and carry an unprecedented amount of information and processing capability in their pockets while enhancing security.

The challenge for all players within the smart card arena will be to work together to provide end users with the appropriate chip technology. This book will hopefully go some way to explain what has been achieved to date and what still needs to be done. For Visa's part, we are proud to be working with member financial institutions and major chip vendors in a global partnership program that will lead to the gradual introduction of smart cards, allowing compatibility among all payment cards worldwide.

Jon Prideaux,
Senior vice-president, New Products, Visa International EU region

Part 1

Background

1

Introduction

The march of the card

Plastic cards are a part of the way of life in most industrialized countries. We use them to identify ourselves, to travel, to gain access to buildings, to obtain cash from our bank, and to pay for goods and services. We are regularly offered new types of cards; many people collect every card they are offered, while others feel that their lives are already excessively controlled by anonymous pieces of plastic.

Nor is this phenomenon restricted to rich western countries. Far Eastern consumers are among the most avid users of plastic payment cards, telephone cards, and travel cards. In parts of Southern Africa, family benefits are paid in cash by a machine, by inserting a card and matching a fingerprint with that stored on record. Electronic purses are used in several countries where the cash economy has all but collapsed. And several cities in China use smart card public telephones, which are equivalent to the most modern in Europe or the United States.

The first plastic cards were little more than a durable business card; the printed information made them more difficult to copy, but it was still read by human operators. *Embossing* made it possible to transcribe the information onto carbon-backed or chemical paper, but the data were still captured by key-punch operators for use in a computer system. *Magnetic stripes* enabled the whole process to be automated to a much greater extent, but the card was and is still essentially an identifier, at most linking the holder to an account. *Signatures* and *photographs* allow the identity of the cardholder to be compared, somewhat imperfectly, with that of the person to whom the card was issued.

Some of today's applications, such as health cards, retail loyalty, and portable data collection, require more data to be stored on the card than a magnetic-stripe card can comfortably handle. But the real reason for using a *chip* in a card, more often than not, is security: smart cards, and the integrated circuits used in them, have several features that allow them not only to store data in a secure way, but also to secure data stored in other computer systems. This book is concerned with these features and their use.

What is a smart card?

Smart cards are also often called *chip cards* or *integrated circuit (IC) cards*.

Throughout this book, we will use the terms *chip card* and *smart card* interchangeably to denote a card meeting the ISO 7810 standard (bank card size and thickness), but incorporating one or more ICs within its thickness (see Figure 1.1). The term *IC card* is identical in meaning, but *chip card* is more easily related to the French *carte à puce* and the German *Chipkarte*. Most of these cards are in fact memory cards rather than microprocessor cards; both types are often referred to as smart cards, but some purists prefer to reserve the term *smart card* for microprocessor cards, whereas *chip card* can include memory cards.

Another term that can cause some confusion is *cardholder*. This does not refer to the person holding the card or even offering it at a point of sale; the cardholder is the person to whom a personal card was issued.

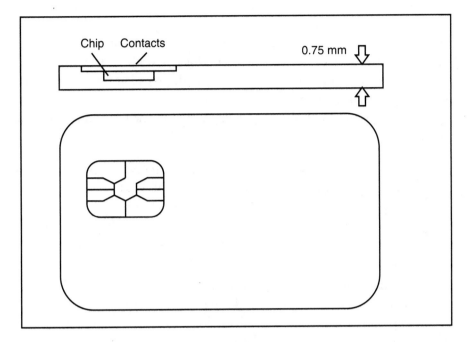

Figure 1.1 Smart card (or chip card).

Systems and procedures

Any security system is only as good as the systems and procedures that use it. Far more burglars exploit an open window or a weak door frame than pick the lock. Most computer system "hacks" involve a user ID and password, which either was divulged to the wrong person or should have been changed or canceled in the meantime.

Many security breaches involving cards do not exploit any flaw in the security of the card itself: Cards are lost or stolen, they are intercepted in the mail, or the fraudster arranges for the card to be delivered to the wrong house. Any proposals to increase the security of card systems must therefore address the whole system and not just the card itself. Using a highly secure cryptographic card badly is rather like fitting the lock to the wrong side of the door.

The security of modern cryptographic systems depends largely on key generation and key-management systems. To continue with the lock analogy, the security of a cylinder lock is not compromised because many people know how it works; but if someone makes a copy of the key or gives it to the wrong person, then the lock is useless. Cryptographic algorithms can be standardized and published, provided that appropriate systems are used to keep the keys secret.

Security is particularly important in smart-card systems for two reasons:

- Many applications are in finance and payments, and potentially quite large sums of money can be involved. Up to 0.2% of the turnover in the major credit and debit card systems around the world—nearly $1 billion a year—is fraudulent. The main aim of using smart cards for these applications is to reduce fraud.

- Many smart-card systems are used in sensitive areas such as personal identification and health. If security is compromised by inadequate systems, the resulting publicity could affect public confidence and reduce the scope for using cards—and indeed any computer storage of data—in these areas.

Market issues

The first smart cards were introduced in 1970 in Japan. The idea was patented in Europe in 1974, and to date over 3 billion smart cards have been sold around the world. This is scarcely a new technology.

On the other hand, there are reasons why smart cards may be about to enter a period of rapid growth:

- The main patents have expired and the field is therefore free for many more players, possibly including some smaller and more innovative companies, to enter the market.

- Concerns about the security of the main competing technology, the magnetic stripe, have increased, and it may now be cheaper to replace the technology than to continue to upgrade the security of magnetic-stripe cards.

- Several applications that are strongly linked to smart cards (such as health cards, digital mobile telephony, electronic purses, and satellite television decoding) are themselves growing rapidly in almost every country.

- The potential market for electronic purses and telephone cards in the former planned economies and in the developing world is estimated at over a billion cards a year.

All of these are reasons for suppliers to invest in and be enthusiastic about the future of smart cards. They are also reasons why it is worth putting time and effort now into understanding the security mechanisms available and into designing systems and procedures that will make full use of these mechanisms.

Organization of this book

The aim of this book is to enable readers to understand the technology of smart cards, particularly those features that relate to security, and the systems and procedures that must be put in place for a smart-card scheme to be effective in protecting data. It should enable a reader to specify the security features required for a scheme in detail and, in many cases, to implement them directly.

It is designed to be read by those responsible for the design, processes, or security of smart-card systems in any application. Some readers will be highly technically aware and familiar with the concepts and terminology; others will be meeting them for the first time. Because some jargon and acronyms are inevitable, we will try to explain any technical terms when we first use them, and there is a glossary at the back of the book.

In the rest of Part 1, we look at the background: the difficulty of defining the problem accurately when faced with a range of misconceptions, prejudices, and sometimes impenetrable computer and encryption jargon. Chapter 3 presents an analytical approach to the task of specifying security requirements and explains the range of tools available for risk management and requirements definition.

Part 2 explores the technology behind smart cards and related products, starting with the basics of card technology, including magnetic-stripe and optical cards as well as IC cards. Chapter 5 is devoted to the science (many would say the art) of encryption and key management.

Identification of the cardholder is a critical problem for any card technology, and often a major role of the card; Chapter 6 looks at the different methods of identifying a person. Chapters 7 and 8 consider the different types of smart cards and the hardware and software elements that make the cards special. Next we look at the other system components, from card readers to computer networks, that can also play a role in enhancing or compromising the security of a card system. The need to protect the card at all stages of its life, from manufacture to disposal, is often overlooked in the early design stages; there must be a positive plan for all these stages, and this aspect is considered in Chapter 10.

Part 3 is concerned with the major applications of smart cards: in telecommunications, finance and retail payments, health care, and personal identification. This part finishes with a chapter looking at the special problems posed by multiapplication cards, particularly where the applications are administered by different parties.

In Part 4 we explore current trends and issues and recommend a security model that can be used for the specification and design of any standard commercial system using smart cards to protect goods and data. Users with exceptionally high security requirements or with unusual usage conditions should adapt the model to their requirements.

The appendix includes a list and brief description of the most important standards in this field, including card, encryption, and procedures standards. These standards are referred to repeatedly throughout the book. The book concludes with a glossary of terms and a bibliography of references and further reading on the cryptographic principles discussed.

2

Problem Definition

Perceptions

Those responsible for introducing plastic cards into our lives have always emphasized the security that they offer. For many years the public believed that cards protected buildings and bank accounts against fraudulent access and that the technology was only accessible to the banks and security companies that issued the cards and administered the schemes.

Up to a point, this was true; cards used for security and financial purposes incorporate very many different security measures, each of which adds to the difficulty of using a card fraudulently or producing a counterfeit card. Unfortunately, familiarity breeds neglect, and most of the measures so carefully put in place are regularly ignored by those who should be checking them. This has led to a great many abuses, which have provided a rich seam for both responsible and sensational journalism.

However, sensational coverage has outweighed responsible reporting, and public perception of the security of card systems may have swung from one extreme to the other. It has been claimed that any criminal armed with a $500 card encoder can manufacture cards that can be used freely in cash machines and retail terminals; this is simply not true. But a grain of truth exists to make the story believable, and the efforts of card issuers to defend the infallibility of their systems have only served to make the lay reader more suspicious.

. . . and reality

The truth is that most abuses stem from breaches of the procedures and checks designed to protect the system. A much smaller number of abuses, often regarded as more pernicious by card-scheme operators, occur because several relevant technologies are now freely available and understood by many competent engineers both within and outside the criminal community.

The first group of abuses (problems with procedures and checks) are often not malicious; they are simply mistakes or shortcuts taken by operators who do not appreciate the consequences of their actions. However, a system that leaves gaps or opportunities for this type of error is defective.

Many of the checks we will describe in Chapter 4 require the operator or cashier to carry out a visual check or to exercise some judgment. These checks are often carried out in a cursory way or ignored completely: cards are handed back before the receipt is signed or the card face is never inspected at all.

Abuses of card systems resulting from intrinsic weaknesses in the technology are, however, growing. As we will see, the most common card technologies are widely available. Many criminals and hackers have sufficient understanding of the systems and coding techniques involved to allow them to produce cards or card transactions that will defeat the weakest links in the chain.

Such attacks can be highly targeted. Some counterfeit cards could only be used in a vending machine or automated teller machine (ATM). Meanwhile, others might not work in an ATM but would pass the simpler checks performed by a point-of-sale terminal or the visual checks made by a cashier with a manual embosser (a "zip-zap"). Many popular forms of fraud involve an accomplice who can be relied upon not to carry out the check. By these means the fraudster produces a bogus financial transaction or gains access to a building or computer system.

Fraudsters have even produced fake ATMs; these look sufficiently realistic for customers to feed in their cards and personal identification numbers (PINs); the details are then captured and used to produce counterfeit cards with known PINs.

Counterfeit cards are most easily used in unattended situations where the appearance of the card is unimportant. International telephones have been particularly targeted using this technique, and one company found that over 40% of its calls to certain African countries were being made with copied or modified cards. (Techniques were swiftly developed to detect suspicious calls and limit their duration.)

The overall cost of these frauds and abuses is much less than many people believe, though. Credit card fraud is about 0.15% of turnover, much less than the cost of most systems currently proposed to eliminate it (none of which would be completely effective). Benefit fraud with card-based systems is also under 0.5%, a small fraction of the fraud experienced by manual systems. Computer systems protected by cards and tokens are very rarely attacked, simply because there are many easier targets. Robert Maxwell, for instance, did not need to break any card security system to defraud his employees of billions of dollars.

Nevertheless, the systems are known to have loopholes, and the risk of a very large loss or a damaging physical security breach is rising continuously. A company that is the target of an attack draws no comfort from the 99 that have had no problems, and every case of fraud helps to undermine the system and public confidence in it.

Calculating the risks: probabilities and odds

Much publicity has been given to stories of factories in the Far East turning out thousands of perfect copies of bank cards and of computer programs that decode cable television signals being freely distributed over the Internet. These stories all have some basis of truth. However, bank card counterfeiting is still more difficult and less prevalent than currency (banknote) counterfeiting, which is estimated to have increased tenfold between 1993 and 1996 as the quality of color copiers increases. The levels of fraud and counterfeiting in the card systems are high enough to be worrying, but not high enough to cause the whole system to collapse; in many countries (including the United States) they are regarded as a normal part of the cost of business.

The common cable television *encryption* systems, many of which have been broken, actually scramble the signal rather than encrypt it with an acknowledged strong algorithm; fast signal processing systems could in principle unscramble the signal without reference to the keys. Presumably this decision was made because it was believed that the extra cost of the set-top boxes for a strong encryption method was not justified.

An important feature of any satisfactory security system is that it not only protects but can be proved to protect. As a minimum, the system must detect any intrusion or breach of security. For "strong" security, it must be possible to analyze every mode of failure and calculate its probability. The system designers must be prepared to stand up in court and defend the system against the evidence of expert witnesses.

No encryption system can guarantee absolute security; no matter how long or complex the key system, there is a minute chance that the first combination an intruder tries will be the correct one. Strong systems trust that would-be intruders also calculate the chances and do not waste their effort gambling on such long odds.

One of the most dangerous assumptions a system designer can make, though, is that all would-be intruders behave "rationally" (i.e., they follow the same thought processes as the designer). Such assumptions are often exploited by hackers and others who approach a problem from a different angle. Criminologists and psychologists can often describe thought processes that are very different from the norm but nonetheless have their own logic. It is often necessary to think laterally or to ask "what if?" There can be many more modes of failure than first meet the eye.

Technical communication obstacles

Another problem frequently encountered by system designers is the difficulty of explaining the problem to end users. One way to accomplish this is to seek a rational, preferably numerical, basis for the system design.

It is common for end users to demand "100% security" or "absolute guarantees" that the system cannot be breached. As we mentioned earlier, this is impossible with any scheme, and it is important to work to realistic criteria. It is often possible to talk in terms of incidents per hundred years or to establish a link between the cost of the security system and the cost of the worst incident that could occur. These concepts are understandable to both parties and often help to break down obstacles.

Computer salespeople often do not help their cause when they use abbreviations and acronyms. The glossary at the end of this book shows the large number of terms used in connection with data security and smart cards. These terms are often used loosely, and customers and salespeople may have slightly different understandings of their meanings. Customers are often unwilling to admit their ignorance of the terminology, and salespeople often repeat the terms used by technicians without fully understanding them.

For security systems in particular, it is important that the criteria are initially expressed in user terms. The translation of these criteria into computer terminology and technical standards should normally follow the first part of the system design.

In Chapter 3, we will consider how these criteria may be defined, analyzed, and verified.

3

Specifying the Requirements

THE FIRST STAGE in any system design is to specify the requirements. This applies to the security aspects of the system just as much as to its functions. In Chapter 2 we mentioned that it is important to specify the requirements in user terms rather than in computer jargon, and wherever possible these requirements should be quantified (expressed in numbers).

In this chapter we consider the aspects of security that we can control, the types and levels of risk a system may face, and the need to balance these against the cost of protection.

Security criteria

Security, even in a data communications system, means different things to different people. There are many ways a system can fail or, even when functioning correctly, can fail to protect its users or their data. Most computer system designs will need to take into account all or some of the following criteria.

15

Safety

Computer systems that protect personal safety normally have a fail-to-safety provision, so that even if the whole system fails nobody is at risk. In some cases, such as *fly-by-wire* aircraft and space travel, this is not possible: without the computer system the aircraft is unstable and uncontrollable. In these cases, voting systems and plausibility checks are applied to all control outputs. It is impossible to eliminate every possible catastrophic failure mode, but the probabilities can be made extremely low—an application of risk management, which we will discuss later in this chapter.

Few smart-card systems will be directly concerned with personal safety, but smart cards may give holders access to dangerous areas or permit them to do dangerous things. It is always important to consider first whether any such risks exist, because the "cost" associated with such a risk is very high.

Nondelivery

This is the risk that data in a communication system is lost. The cost of nondelivery often depends on whether we can detect the error. If we receive message number 133 followed by 135, then we can request a retransmission of message 134. If we do not know that it was missed, the cost might be an order or a transaction that we failed to transmit to the bank. In private and local networks, it is nearly always possible to detect nondelivery; in public networks, the risk of it passing undetected is much greater, and systems should build in message numbering or other checks.

Smart-card systems are often concerned with recording transactions or events. It is easy to lose a single event unless they are numbered; numbering also helps check against duplication. Another form of cross-check is double entry; we may, for example, both keep a tally of the total amount spent on each category of goods and record the individual transaction amounts. Use of such checks will help detect errors, but we must still have some way of recovering from them and, if possible, reinstating the lost transaction.

Accuracy

Here we are concerned with the possibility that data may be recorded or transmitted incorrectly. This normally occurs through random errors (perhaps caused by a poor contact or electrical interference). Smart-card systems must often have very wide electrical tolerance ranges to deal with the many different devices and contact conditions they may encounter; these increase the risk of errors passing through. In practice, the card itself, and its interface with the terminal, are rarely the source of these problems; they are much more often caused by the external systems. But the system design must take into account all possible sources of errors.

The cost of such an error will depend on the type of data: A single character error in a text string may be easily spotted and not important, while a one-digit error in a transaction amount, or a single-bit flag, may be much more difficult to detect and will have a bigger effect. Systems can make use of plausibility checks to help detect serious errors.

As well as the parity and *cyclic redundancy checks* (CRCs) that are widely used in computing and data communications, security-conscious smart-card systems make use of *message authentication checks* (MACs) on messages and critical data fields. These techniques will be discussed in more detail in Chapter 5.

Data integrity

Data stored on a card or another computer system should be protected against alteration, whether malicious or accidental. This is an area in which some types of smart cards excel: Areas of memory can be protected against access by unauthorized applications or individuals, and some systems make use of memory that can be written only once, such as memory *write once, read many times* (WORM) or *flash* memory, which after being written can only be erased as a block.

Malicious attacks on the integrity of stored data are exceptionally serious when they involve encryption keys, expiration dates, or authorization codes. In these cases the whole purpose of the card system may be perverted. Designers should normally assume that *malicious attacks will occur;* the system must be capable of maintaining its integrity in the face of such attacks.

It is also important to ensure that transaction details are not altered during transmission to the host system. MACs, digital signatures, and other forms of encryption can again be employed and are more secure than other forms of checksum or CRC.

Confidentiality

Many smart-card systems are concerned with protecting the confidentiality or privacy of information. This applies not only to the main data files stored on the card, which might include medical details or credit limits, but also to data stored on other computer systems that may be accessed by using the card, such as the balance on an account or the details of a company's customers, orders, or designs.

It is important to take a realistic view of the risk involved here: Many companies spend a fortune protecting the confidentiality of internal information and a similar fortune publicizing the same information! Even with slightly more sensitive information, the risk may be minimized by ensuring that only part of the information is stored on the card or transmitted at any one time; codes rather than text may be used to describe items. The risk should be quantified where possible: How much could the worst-case incident cost the company?

Some confidentiality requirements are imposed by law: in the case of payments or medical records, the cardholder is entitled to privacy, and the card issuer or scheme operator should ensure that the best current practice is followed. For small value payments, the card may form part of the wall of privacy: Data protection officials in some countries have ruled that details of such payments (such as location or goods purchased) should not even be stored by the card issuer. In this case the card is the only place where they can be stored anonymously.

There is a big difference between malicious breaches of confidentiality (*eavesdropping*) and accidental ones (often called *leakage*). Leakage is usually caused by a system malfunction, whereas eavesdropping implies a weakness in the system design (a *systematic error*), which may be exploited over a period of time to maximize the damage caused.

Access controls and encryption are the tools most often used to *protect* confidentiality and privacy; it is usually more difficult to *detect* a breach of privacy.

Impersonation

Impersonation, or *masquerade,* is the risk that an unauthorized person (someone other than the cardholder) can make use of the functions allowed by the card. Many smart cards perform a personal identification role; in this case we are looking at the chance of others being able to pass themselves off as cardholders. We will consider this in more detail in Chapter 9, when we look at personal identification techniques, but from a risk point of view we must be able to assess and quantify the cost associated with a successful impersonation.

This is an area where it is particularly important to look at the whole system and not just at the card itself. Impersonation incidents can have a particularly damaging effect on public confidence in a system.

Repudiation

A scheme operator must often be able to prove that a particular transaction took place, that it was authorized in the correct way, and that it has not been altered subsequently. Neither the cardholder nor the merchant can then repudiate the transaction or claim that it never took place.

Digital signatures, using public key cryptography, are the answer to this requirement; they are often referred to as *transaction certificates* and may be stored by the scheme operator along with the transaction details.

The system designer must weigh the cost of including in the system an automatic method for handling demands for proof of transaction (verifying transaction certificates) against the likely frequency of such demands. The very existence of a stored transaction certificate will deter many spurious claims. Support functions such as this account for many of the hidden costs of any card management system; smart-card systems usually aim to automate as many of them as possible, but they cannot be eliminated completely.

Quantifying the threat

Possible outcomes and costs

We must now examine the system itself and list, in user terms, the possible outcomes of any of the security problems listed in the previous section.

For example, a breach of confidentiality could result in:

- The cost of an investigation and writing apologetic letters;
- The loss of a customer;
- Compensation;
- A lawsuit;
- Widespread loss of confidence in the system and loss of customers.

These are associated with increasing costs per incident, but probably also with decreasing likelihood. If the system has been designed to support customer service inquiries and incident investigation, then the damage can probably be minimized.

In a card payment system, nondelivery incidents could range from loss of a single transaction to a whole day's trade of one merchant. More widespread loss of data from a single incident can probably be discounted except in the context of a disaster scenario (such as the computer center being destroyed by fire). We said earlier that almost all nondelivery incidents can be detected, so the cost associated with this type of incident is that of recovering the data. Taking into account the average value of transactions, it could be decided that the cost of recovering a single transaction is higher than the likely loss—although the perception of system integrity must also be taken into account.

Where it is difficult to assign a monetary value to each of the outcomes, we can use a grading scale, for example:

- *Grade I:* Significant danger of death or injury;
- *Grade II:* Likely to endanger the survival of the company or business;

- *Grade III:* Likely to incur significant loss of business, external costs, or compensation;
- *Grade IV:* Likely to incur significant extra work or delays;
- *Grade V:* Nuisance value; small delays or extra processing requirements.

Objects threatened

Having considered the possible outcomes from the various types of security breaches, it is useful to categorize these according to the object that is being threatened. This also acts as a check to ensure that all of the outcomes have been considered.

Usually, the object most at risk is an item or set of data. This may be a single field in a personal or account record, a complete record, or the whole file. The most serious incidents involve key fields and access rights (computer hackers like to set up or use system administrator accounts). Fields or items that have broadly the same characteristics and are subject to the same risks can be grouped together.

Hardware can also be the subject of a threat: What if the card itself is lost, stolen, or damaged? A PC or disk drive that holds data can be damaged or stolen, or the modem that gives access to the data may malfunction.

Again, the most serious problems are when the hardware item forms part of the encryption mechanism. It is increasingly common to encrypt all files on a disk (for reasons of both space and security); many backup systems have problems with compressed and encrypted files, and whole volumes are sometimes lost—particularly where the backup was incomplete.

Causes and modes of failure

As well as listing the possible outcomes and estimating a cost associated with each, it is worth considering the ways in which each of these outcomes might arise. Specifying security requirements is usually an iterative process; the first time we do it we may have relatively little idea what the system architecture will look like or what its components will be. As the design progresses, it is easier to imagine ways in which the system could fail and what its weaknesses are. Trying to list causes of failure before the system has been designed, however, has the advantage of there being no preconceptions. And nobody can be offended because their design is being criticized!

Security incidents are most often caused by:

- Hardware failures (whether permanent or transient);
- Software errors (usually caused by inadequate specification or testing);
- Network and communications problems;
- Procedural errors;
- Malicious attacks (often exploiting a failure in one of the previous categories).

Causes of failure can be categorized as:

- Random: Single failures that cannot be predicted, although it may be possible to assign a *mean time between incidents* (MTBI) or other probability measure to them;
- Intermittent: Incipient failures can often cause a large number of incidents and incur large costs. Statistical analysis is the key to detecting intermittent failures and must form part of the overall system design if it is to be useful when required. This is particularly necessary for the large distributed systems typical of smart-card schemes.
- Systematic: Failures that will occur every time a particular combination of events or a particular sequence of transactions occurs. These are often system design or software faults, and they are common in new systems and software versions. They can be reduced (but never completely eliminated) by careful specification and testing.
- Serial: Where one failure is likely to lead to another. Maintenance and backup systems are often less secure than the normal systems, which is the opposite of the desirable situation.
- Catastrophic: Failure of the whole system or a large part of it. The reasons for this may be internal to the system (e.g.,

corruption of a database) or external (e.g., a fire or electrical failure). At this stage we are mostly concerned with security and cost implications; we need to establish the cost of a security failure in order to make decisions as to whether we need, for example, a second site rather than a backup database or an un-interruptible power supply (UPS). Disaster planning is a separate issue, concerned with keeping the whole business going.

Frequency of incidents

After examining the potential sources of failure and assigning a cost to them, the next stage when specifying the requirements is to specify a maximum frequency for each type of failure. Because of negative psychological effects on staff and customers, a maximum frequency should be given even for failures that have only a nuisance value.

The likelihood of failure can be expressed as a probability (e.g., 1% per year). But a more easily understood measure is usually the MTBI—this is the inverse of the probability, so 1% per year is an MTBI of 100 years. The result is Table 3.1.

Table 3.1   Sample Requirements Specification

Failure Outcome	Possible Cause	Minimum MTBI (Years)	Per (Unit)
Customer unable to use card in shop	Card failure	3	Cardholder
Customer unable to use card in shop	Terminal failure	1	Terminal
Customer unable to use card in shop	System failure	1	System
Scheme unable to collect data from terminal	Terminal failure	1	Terminal
Scheme unable to collect data from terminal	System failure	1	System
Customer account error	Card error	100	Cardholder
Customer account error	Terminal error	100	Cardholder
Customer account error	Comms error	100	Cardholder
Customer account error	System error	10	System

Risk management

We have seen that many card-based systems suffer from imperfect security. One reaction to this might be to insist on the best available technology in all areas. This approach is valid in some cases, but can result in an inappropriate use of resources. In the case of smart cards, it would result in military-grade cryptographic processors, very large segmented-memory requirements, and a card cost outside the pocket of most schemes.

The security measures must be matched to the application, and the way to do this is to repeat the process described earlier for the specification of the requirements. This can only be done after an outline design for the system has been completed.

The designer will compare the security-requirements specification with the actual likelihood of failure of the elements concerned. In the early stages of the design, this may only be an estimate; if a computer model is being used, then different values may be used to find which areas are particularly sensitive—it may be worth looking at these areas in more detail or specifying higher grade products.

For hardware items, MTBI figures are often available from the manufacturers (for any security-conscious system, they must be). They are usually obtained by a combination of calculation and testing. MTBIs are usually lower than mean times between failures (MTBFs) by a factor of two or more. For a large system, even quite long MTBIs may yield significant incident rates: a population of 100,000 cards with an MTBI of five years will yield over 50 incidents a day. Card life is discussed further in Chapter 8.

The result of this process is a table showing the probability value of each of the events considered (Table 3.2).

The first thing this does is to concentrate the mind on the areas that have relatively high probability values: these are areas where potential disasters loom and where additional effort is most likely to be well spent. In the case considered here, card failure is the largest single cost; increasing card life may be a very expensive proposition, but the operator could consider offering customers special wallets or insisting on a higher specification plastic.

Table 3.2 Probability Value Table

Failure Type	Cost	Probability (Per Unit Per Year)	Number of Units	Probability-Value
Single card failure	$50	0.2	100,000	$1,000,000
Terminal failure	$500	0.3	1,000	$150,000
System failure	$10,000	1	1	$10,000
Random card error	$1,000	0.001	100,000	$100,000
Random terminal error	$5,000	0.05	1,000	$250,000
Card programming error	$5,000,000	0.001	1	$5,000
Terminal programming error	$250,000	0.01	1	$2,500
Operator private key discovered	$50,000,000	0.00001	1	$500

The next stage is to look at the cost of possible countermeasures. A proper risk analysis should ensure that if a countermeasure reduces the probability value by more than the cost of the countermeasure, it will be adopted. In the case of a new hardware or software design, it is often difficult to determine in advance just how much the countermeasure will cost: We may study an area for weeks and conclude that there are no viable improvements. Judgments must be made.

When the process is complete, we have a list of the risks and probabilities that can be presented to senior management. The business may need to seek ways of protecting itself against some of the more serious effects.

Standards

Use of standards within specifications

Standards offer a shorthand way of specifying a group of requirements, and in some cases an independent laboratory can test equipment or software against the standard, providing a certificate of "type approval" that prevents users from having to check every aspect of the item's operation.

There is a very large number of standards applicable to smart cards in one way or another. The appendix lists around 40 of them, and this is by no means a comprehensive list. Some of them define interfaces or procedures, while others set minimum requirements for physical or reliability aspects. In very few cases, however, is there provision for certifying products against the standard.

Standards are often a statement of best current practice, and in that context should be followed or specified wherever possible. Almost all commercial encryption packages follow the relevant standards very closely. Specifying long lists of standards to which a product must adhere is not necessarily helpful, however, and may obscure the more important requirements of the individual case.

In some cases, such as the ISO 7816 part 1 physical standard for smart cards, the standard is now regarded as an absolute minimum, and most manufacturers have set their own standards well in excess of this level.

Classes of security

There are two groups of standards that grade security products and systems according to the level of security they aim to provide and set a mechanism for testing products against their stated aims. They are:

- The Trusted Computer System Evaluation Criteria (TCSEC), used in the United States;
- The Information Technology Security Evaluation and Certification scheme (ITSEC), used in Europe.

The two standards are comparable but not identical; TCSEC grades products from D (the lowest level) to A1 (the highest) according to the level of security they are trying to achieve and the extent to which they are able to demonstrate that they have achieved it. A suitably designed smart-card system should be able to meet level C2 (controlled access protection); to reach the higher B1–B3 levels, formal design and analysis procedures must normally be used.

ITSEC gradings cover a wider range of security objectives, grouped under the headings:

- Identification/authentication;
- Access control;
- Accountability;
- Audit;
- Object reuse;
- Accuracy;
- Reliability of service;
- Data exchange.

A supplier may define a product's objectives by using any combination of these criteria, but common requirements are grouped together in predefined classes. For smart-card systems, the classes F-DI (high standards of data integrity during data exchange), F-DC (high standards of confidentiality during data exchange) and F-DX (combines the requirements of F-DI and F-DC) are likely to be the most important.

Under ITSEC, products and their documentation are assessed by a commercial licensed evaluation facility (CLEF) using criteria for correctness and effectiveness; they are given a rating from E0 (inadequate assurance) to E6 (the highest level). For ratings of E4 and above, formal models and design processes must be used; some smart cards have reached level E3, but it is unlikely that a complete system could be certified at this level.

Both ITSEC and TCSEC are likely to be eventually superseded by the common criteria for computer system security, currently being developed by ISO.

Quality assurance

Increasing importance is being placed on quality assurance (QA) as a means of ensuring the correctness of an operation. Many government departments in particular will only buy from suppliers that meet the International Standards Organization (ISO) 9001 or 9002 standards, or their national equivalents. ISO 9000 manufacturers face increased costs when they buy from noncertified suppliers, so the pressure continues downwards.

As with the functional standards mentioned earlier, the standards themselves should be followed insofar as they represent best current practice. The ISO 9000 standards enforce a high standard of record keeping, which is very important for a card issuing or customer help-desk operation. Many organizations, particularly smaller companies, have found, however, that obtaining and maintaining full certification against these standards represents an excessive overhead cost.

Documenting the specification

Initial system specification

The initial security specification for a system must include:

- A list of the security objectives (what is the system trying to achieve?);
- A maximum frequency for each of the outcomes considered, similar to Table 3.1.

This "bare bones" specification, which may be a single section within the overall functional specification, will form the basis of a system design.

Analysis and iteration

The design should then be analyzed and the risks associated with each element listed and quantified (as in Table 3.2). These can then be modeled with a simple spreadsheet—a process to which we will come back in Chapter 18—to find areas that are particularly sensitive or where expenditure could most easily be justified. Following redesign, the process is repeated until the overall system objectives are met and any weak areas have been covered.

There is a danger in relying too much on mathematical models or spreadsheets in a process such as this: It is easy to overlook a whole category of threat or to make unjustified assumptions about people's behavior. Any formal analysis should therefore be backed up with verbal explanation and a statement of how the system objectives are met. It is often helpful to describe how the protection mechanisms would work in the face of:

- Random errors, for example in communication systems;
- Hardware failures in any system component;
- Software errors;
- Malicious attempts to break the system, based on any of the criteria listed earlier in this chapter.

If the analysis holds for each of these cases, then it is likely that the structure of the system is sound. The descriptions should form part of the security specification. For a payment system or other system in the public domain, they should be carefully written, as they could be required if an attack reaches the courts and evidence of the system's soundness is needed.

Component security objectives

For a larger system, the process of specification must then be repeated for each of the components of the system. Where the top level specification makes assumptions about the security of a component, these assumptions should form part of the security objectives for that component. If its designers find these unrealistic or unachievable, then the system analysis must be repeated with the new data.

Assumptions about the security offered by an individual component must also take into account the type of user involved, the application environment, and the way the component is being used in the system. The effectiveness of every tool and technique depends on its context. In Part 2, we examine the tools available and the contexts in which they can be used.

Part 2

Technology

4

Card Technology

Visual features

Cards have been used for identification purposes for centuries. Diners Club issued the first plastic cards (what we would now call a "travel and entertainment" card) in 1950; the first plastic credit cards were issued by Bank of America in 1960. These early cards were made of a single layer of polyvinyl chloride (PVC) printed with a very simple pattern. The customer's name and account number was printed on the card using raised lettering, or *embossing*, in gold or a contrast color, a process known as *tipping*. At the time, color printing was only generally available using rotary printing machines, which could not be used with the relatively rigid plastic cards. The addition of the embossing and tipping made sure that only plants built specially for this purpose could produce these cards, and because the cards were rare and considered to be of high value, people accepting them checked the signature carefully.

Figure 4.1   Microprinting. (Photograph courtesy of Giesecke &
Devrient GmbH, Munich).

Today, thermal transfer printers made specially for printing bank-sized cards will fit on a desktop and can be bought for less than $5,000. Manual embossing machines are even cheaper. The original designs can be scanned and copied by any desktop publishing system. Card issuers therefore make use of a very fine printing process known as *microprinting* and complex patterns and color changes, which are designed to be distinctive and difficult to reproduce. Printing techniques developed for banknotes, such as *guilloches* (finely printed spirals) and rainbow printing (graduated color changes), are useful because they defeat the resolution of color photocopying or scanning. It is worth looking at some of the characteristics of a modern card, particularly a bank "gold card" or a pass from a security-conscious establishment, under a magnifying glass to see some of its special features.

In many retail situations, however, assistants do not have the time or the inclination to scrutinize cards this thoroughly (any more than they scrutinize bank notes). And there are too many designs in circulation for them to be able to know the distinguishing features of each. Holograms are used to provide a quickly assimilated check of the authenticity of cards belonging to one of the major payment schemes: Visa's dove and Mastercard's interlocking globes can be recognized quickly and are very difficult to reproduce accurately. On some cards, these same patterns are printed on the card with ink that is only visible under ultraviolet light.

Other visual features used on cards include the signature and/or the photograph of the cardholder. Either or both of these can be printed onto the card as a part of the personalization process rather than being added onto a paper signature strip. Here the difficulty is not only the thoroughness with which the signature or photograph is checked, but also that both characteristics often appear quite different in the small space allotted to them on a card from that in real life. Both signatures and appearance also change with time, and they depend on a highly subjective assessment on the part of an untrained observer. Something better is needed if the card is to be a reliable form of identification.

Magnetic stripe

The most common form of card technology for automatic reading is the magnetic stripe. A half-inch-wide (12.7-mm) strip of magnetic tape is bonded to the card substrate. The individual magnetic particles are aligned along the stripe, but when unencoded, the individual particles may be magnetized either from left to right or from right to left, so there is no overall polarization. Blank white cards with a magnetic stripe can be freely purchased from card manufacturers and dealers.

Encoding and decoding

To encode the stripe, a magnetic field is applied using a coil-wound magnetic head; this results in all the particles being polarized the same way. When the encoding current is reversed, the polarization also reverses. Figure 4.2 shows the principle.

The stripe is read by passing the card in front of a reading head; each polarization reversal results in a pulse of current in the coil, positive or negative depending on the direction of the reversal, as shown in Figure 4.3. The faster the card is read, the bigger the pulse.

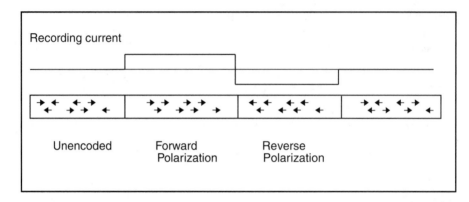

Figure 4.2 Magnetic-stripe encoding.

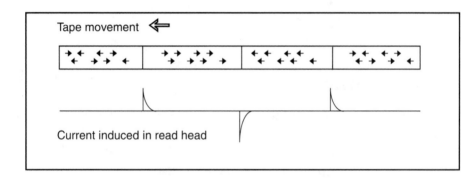

Figure 4.3 Magnetic-stripe reading.

These pulses must then be decoded into 0s and 1s: Several different encoding schemes are possible, but the most common is known as F/2F. In this scheme, a 0 and a 1 are the same length (e.g., one-75th of an inch): a 0 is a single *domain* with no flux reversal, while a 1 has a flux reversal in the middle (see Figure 4.4). As with most magnetic-stripe reading, accurate decoding depends on being able to distinguish between the two lengths; it is therefore much easier when the card is being read by a motorized reader (which delivers a fairly uniform speed) rather than by a manual reader (of the swipe or insertion type). Poor encoding can make reading very difficult: If the speed of travel of the card through the

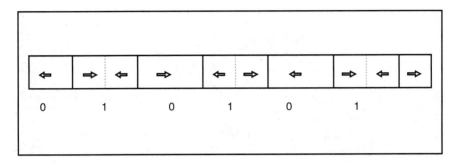

Figure 4.4 F/2F encoding.

encoder varies, then the bit length is also likely to vary, and even the level of magnetization will often be dependent on the speed. Some encoders may also show some "spillage" of data from one track onto the adjacent tracks.

Decoding of F/2F data is usually done by a single IC. Despite the problems referred to in the previous paragraph, a good decoder chip will read almost any card first time, provided that the stripe has not been physically damaged. Less satisfactory decoding chips will only read cards which are very close to the specification.

Copying and counterfeiting

Magnetic-stripe technology is essentially the same as that of a tape recorder. Most bank cards and identification cards conform to one single standard (ISO 7811—see the appendix). This specifies the position of the data on the card, using three tracks and the F/2F encoding scheme. It is therefore relatively easy to construct a card reader from components, and in fact complete card readers with serial or keyboard interfaces are available from a wide variety of sources (see Figure 4.5).

Building an encoder is slightly more difficult, first because of the need to control speeds and tolerances quite accurately, but also because there are no standard components for conversion of data into the encoding current required to drive the head. However, it is a straightforward engineering task for a suitably equipped organization.

Because magnetic-stripe cards are used in such a wide range of applications, many organizations have a valid need to encode cards,

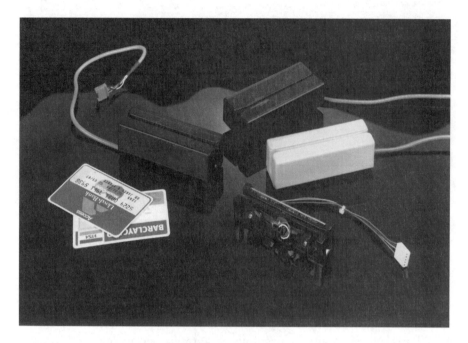

Figure 4.5   Magnetic-stripe card reader. (Photograph courtesy of Datastripe Limited, Camberley, England).

and several encoders are commercially available at prices well below $1,000 (see Figure 4.6). Reputable encoders will usually reject requests to encode bank cards (unless this has been disabled; for example, when a machine is sold to a bank), but there are many other security applications of cards for which even this limited protection is not available. Designers of magnetic-stripe card schemes should always assume that cards can be reencoded or cards with identical encoding manufactured.

The technique known as *skimming* involves reading the data from a valid card and copying it onto another card. If the card is physically a copy of the original, then it may pass a visual inspection as well, but even white card copies may be used in many unattended applications.

Many security-conscious card schemes do incorporate techniques to reduce the risk of copying: Check digits and other algorithms can prevent new numbers from being generated, while ATMs and similar devices can leave tell-tale marks that are not copied by standard techniques on the

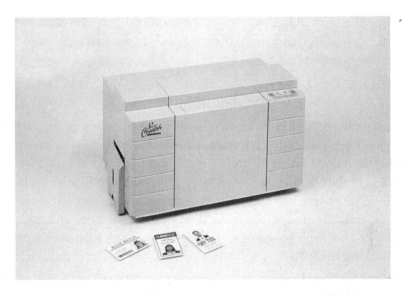

Figure 4.6 Desktop card printer/encoder. (Photograph courtesy of Fargo Electronics, Inc.)

card. By these and other means, as we will see in Chapter 12, the main bank card schemes have controlled the spread of fraud.

High-coercivity cards

The data on standard magnetic-stripe cards is often erased or corrupted by stray magnetic fields. This problem can be reduced considerably by using *high-coercivity* (HiCo) magnetic material. HiCo stripes can only be altered by applying a magnetic field stronger than most common permanent magnets.

HiCo cards also offer a higher level of security against copying. Encoding HiCo material requires more energy (higher recording currents), and it is difficult to ensure that data from one track does not spread into the adjoining track. Once encoded, a HiCo card is no more difficult to read than a LoCo card. But from a commercial security standpoint, the main advantage is that few commercial encoders can handle HiCo stripes.

The standard for HiCo encoding, ISO 7811-6, has only recently been introduced and is not widely used, so there is also scope for using different encoding schemes. In this case, someone wanting to encode cards would have to work out the encoding scheme as well as manufacturing a special device for the purpose.

Other magnetic card types

The magnetic stripe may be located in different positions on the card or it may be discontinuous. The magnetic material may be applied with a printing technique rather than as a tape, and this further increases the flexibility of the operator to locate data in different positions on the card.

These techniques will appeal to operators whose card schemes are large or critical enough to justify the expense of nonstandard card production, but which are not so widespread or offer such large returns that fraudsters will be tempted to go to similar lengths to duplicate the nonstandard cards. Japanese telephone cards, which have one side of the card covered in magnetic material, can be used as a basis for manufacturing counterfeit cards of this type.

Enhancing security using complementary technologies

Several methods can be used to improve the security of magnetic-stripe cards. Those that have been tried most extensively are:

- *Watermark tape:* A thin strip of magnetic tape, with the particles aligned across rather than along the tape, is bonded to the top of the card, on the same side as but above the magnetic tape. The watermark tape is read by special readers, which can be fitted to ATMs and in other critical locations. The tape contains a security code that is unique to every card; the code, or data algorithmically derived from it, is also contained in the magnetic stripe. If the stripe and watermark do not correspond, the card is judged to be counterfeit. This system depends on the proprietary nature of the tape manufacturing process; unfortunately, it also means that there is ultimately one source for the readers, whereas one of the attractions of magnetic stripe technology is the wide variety of reader makers. In trials by the major card schemes, reliability was also lower than would be needed for a full rollout.

- *Holomagnetics:* The concept here is similar; it involves using a machine-readable hologram in place of the visual hologram. Again, this is only suitable for special locations and types of reader; it could be used for protecting ATMs but not point-of-sale readers.

- *Card signatures:* Small variations in the quality of the magnetic tape or in the length of individual bits can be measured by specially equipped card readers. A "signature" made from these variations is stored, in encrypted form, on the standard tracks. These systems should detect not only counterfeit cards, but also unauthorized attempts to alter the data on the card. There are at least two proprietary systems (both from U.S. manufacturers) using these techniques, but they are not widely used in live systems. This is probably because magnetic stripes are always prone to changes caused by wear and magnetized heads in readers; such wear would lead to cards being falsely rejected.

Optical

Optical cards, using a laser reader detecting the depth of a reflecting surface on the card in the same way as a compact disc (CD), can be used to store a very large amount of data: the most common proprietary standard stores 6 MB on one side of a standard-sized card. The technology has been available since the middle of the 1980s. In the early 1990s, it was licensed by Canon and other manufacturers, which have developed commercial products suitable for reading the cards (see Figure 4.7).

The technology appears to be reliable, although because of the limited number of cards in use and the small pool of manufacturers, both cards and readers are still rather expensive. The quantity of data stored in nonvolatile form on a card makes it attractive for applications such as health records and hospital data. The high cost and limited supply of readers ensures a reasonable level of protection for the data.

It would be difficult, however, to prove the confidentiality of the data or to police access to it, particularly if these systems became more

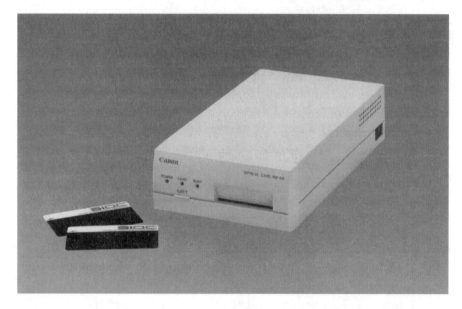

Figure 4.7 Optical card and reader. (Photograph courtesy of Canon Europa N.V.)

common. Data on an optical card may be stored in an encrypted form, but in that case the responsibility for key storage and comparison would rest with the system, thus removing the advantage of complete portability from the card.

Current generation readers rely on software to control access to the data on the card; this is not really adequate for full-scale live applications on such sensitive data.

Smart cards

Origins and development

In 1970 a Japanese inventor, Kunitaka Arimura, filed the first patent for what we would now call a smart card. His patent was restricted to Japan and to the technical aspects of the invention. Japanese cards are manufactured under an Arimura license.

Between 1974 and 1976 Roland Moréno in France successfully patented several functional aspects of the smart card and sold licenses to

Bull and other manufacturers. Bull further developed the microprocessor aspects of the smart card and now holds several patents in its turn on this aspect of the technology. Moréno's company, Innovatron, has pursued an aggressive policy of licensing and litigation worldwide, which has restricted the number of companies working in this field. However, the most important patents expired early in 1996, so the number of system and suppliers may be expected to increase in the next few years.

The most important aspect of the smart card, as Moréno recognized, was the control of access to the card's memory through the use of passwords and other internal mechanisms. From a technology point of view, it is important that parts of the memory are only accessible after certain specific operations have been performed. This makes the chip logic more complex, but dramatically simplifies the work of the surrounding system elements, particularly in the areas of encryption and key management.

Most smart cards, as we saw in Chapter 1, do not contain a microprocessor, but only memory governed by some fixed logic. The volume market, and the commercial success of smart card manufacturers, has depended on telephone cards and other simpler cards. The market for all card types grew very slowly to begin with; by 1989, volumes were still less than 50 million cards a year. In the 1990s, however, growth has accelerated sharply, and by 1995 the volume of memory cards sold exceeded 300 million, with microprocessor cards less than a tenth of that figure (see Figure 4.8). Further rapid growth is forecast as new markets and applications are opened up. Nevertheless, these figures are small in comparison with the number of magnetic-stripe cards manufactured in the world each year.

Elements of the technology

As we will see in Chapter 5, smart cards are usually the same shape and size as the other types of card mentioned earlier, although there are also other possibilities. The chip is contained within the thickness of the card, so one side is usually free for printing graphics, including microprinting and any of the other visual security features. But the security of the card resides in the chip itself, its manufacturing methods, and the data contained within it.

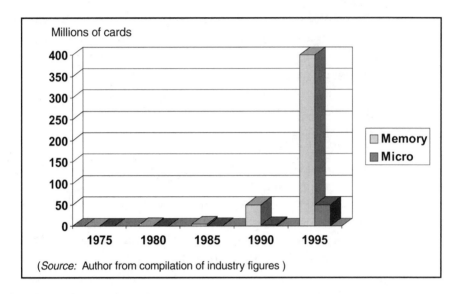

Figure 4.8   Growth of smart-card sales, 1975–1995.

The card consists of a plastic carrier, in which is embedded a specially designed integrated circuit and either a set of contacts or an aerial for contactless operation. Other important elements of the system are the readers and computer systems that will use the card in operation, as well as the systems for manufacturing, issuing, and controlling the card itself and the keys that it contains. These elements are all described in more detail in Chapters 8 and 9.

Standards

A glance at the appendix shows the profusion of standards applicable, in one way or another, to smart cards. Each of these standards will be mentioned in the more detailed sections later in the book. However it is important to mention the several families of smart-card standards that have developed independently.

The most important of these are:

- *ISO 7816:* This set of standards is a logical development from ISO 7810–7813, which cover most current magnetic-stripe cards used in banking and financial applications. ISO 7816 defines a contact card containing a microprocessor that can be used as a direct replacement for a bank card. It is a relatively low-level specification and does little to define the functions of the card.

- *EMV:* Between 1993 and 1996 the major credit-card schemes (Europay, Mastercard, and Visa) developed a further set of specifications based on ISO 7816, but covering the core functions of a bank card in much more detail.
- *ETSI:* The European Telecommunications Standards Institute has been responsible for a set of standards that cover smart cards for use in public and cellular telephone systems.

This still leaves a large majority of all smart cards not covered by any internationally recognized standard. Memory cards in general use a manufacturer's standard, while most smart bank cards currently in issue were designed before the EMV standards were published. Confusion about smart-card standards, and the absence of standards in many areas, have contributed to manufacturers' reluctance to design readers and terminal equipment for general use.

Many of the standards include such a wide range of options that they are almost meaningless: for example, there are at least ten competing electronic purse standards in Europe, all of which meet the draft CEN standard for electronic purses, but none of which is compatible with any other. This situation is improving as de facto standards emerge in each industry, but many manufacturers are still wary of making too great a commitment to any one set of standards.

Hybrids

One card can make use of several technologies. In particular, many smart cards will also carry a magnetic stripe; this applies particularly to bank cards designed for international use.

When a hybrid card is used, there must be rules governing the priority sequence for trying the different technologies: usually the chip must be tried first if the reader is able to do so. If there is no chip, or if the chip is unserviceable, then the reader *may* be allowed to use the magnetic stripe. The application must be aware, though, that the magnetic stripe does not incorporate the security mechanisms that are built in to the chip, so extra precautions must be taken to prevent fraudsters from destroying the chip (by applying a severe overvoltage, or simply by smashing it with a hammer) and gaining access to the account by using the less secure technology.

PCMCIA cards

The growth of the market for laptop computers brought with it the need for interface cards in a very small and standardized size, distinct from the PC bus. This demand was echoed by the makers of scientific instruments and telemetry equipment, who needed not only interfaces but also portable memory.

The Personal Computer Memory Card Industry Association (PCMCIA) has produced a set of three standards, which allow not only small memory cards but also devices such as disk drives to be incorporated into a card format: slightly longer than a credit card, and up to 10.5mm thick in the case of disk drives. PCMCIA memory cards can contain any type of semiconductor memory (it is usually one single type rather than a mixture), and many of them include a battery to sustain the memory. Memory sizes using this technology can be many megabytes.

There is very little competition between PCMCIA cards and smart cards: They each have their applications and there is little crossover between them. PCMCIA memory cards are used for transferring data and programs between laptops or instruments and a desktop computer system. No PCMCIA card, as far as we are aware, has the controlled access to memory that is the special feature of the smart card, although it would be completely practical to make a card in a PCMCIA format with a smart-card chip inside it. Users of PCMCIA cards must normally make use of one of the standard PC file encryption packages and enhanced password systems to protect their data.

Others

Other technologies are used in specific sectors or applications (e.g., radio isotopes) and proprietary units such as the button memory chip. Two more important technologies must be mentioned, however; they are both commonly used for access control and other automatic identification applications, and each has some relevance to the smart-card market.

Barcoding

Barcodes are now very widespread, particularly in the form of the article codes on consumer goods (UPC in North America, EAN in Europe and elsewhere) and in industrial control applications.

Barcodes are highly standardized; they can be produced very cheaply (cards or labels can be printed on site), and there is a wide variety of reader types available. The ease with which they can be produced, however, makes them inherently unsuitable for security applications: a card can be photocopied or labels printed on almost any computer printer. Infrared wavelengths can be used to give some elementary protection, but even then the codes can be reproduced using many common word-processing or printer-control packages.

In the end, however, the limitation of barcode technology is that it is read-only: It is only capable of delivering an identification number. The authentication of the medium and of the cardholder must be carried out by another method.

Barcodes are often used as a supplementary control during manufacturing and card issuing, as they can be read before the card is personalized.

Radio frequency identification (RFID)

RFID is used in access control, traffic control, and some specialized industrial control applications. The tag, which can have many shapes, including that of a credit card, contains an antenna. When the antenna comes within range of a reader, it generates enough energy to power up a circuit in the tag and to transmit the identification number contained in the tag.

Although RFID tags are relatively difficult to copy, the limitation of the technology is again that it is read-only. There are variants of the RFID tag that allow writing, but these are usually very large and expensive and demand high power. The only version that overcomes these problems is the contactless smart card, which will be covered in Chapter 7.

5

Encryption

Cryptology overview and terminology

Cryptology is the science of codes and ciphers. It is an ancient art and has been used for many centuries to protect messages sent between military commanders, spies, lovers, and others who wished to keep their messages secret.

When we are dealing with data security, we need to prove the identity of the person sending or receiving the message and to show that the message contents have not been altered. These three requirements, for confidentiality, authentication, and integrity, are at the heart of modern data communications security, and they can all be addressed by some form of cryptology.

Cryptology has its own jargon and acronyms. A fuller understanding also requires some knowledge of mathematics. Readers who would prefer to avoid the explanation but still understand the main issues should skip to the summary at the end of this chapter.

To protect our original data (known as the *plaintext*), we encrypt it using a *key,* so that the data cannot be understood by anyone reading it. The encrypted data (known as the *ciphertext*) appears to be a meaningless series of bits with no clear relationship to the original. To restore the plaintext, the receiving party decrypts it. A third party (such as an attacker) can use *cryptanalysis* to try to restore the plaintext without knowing the key. It is important to remember the existence of this third party!

Encryption has two main components, an *algorithm* and a *key*. The algorithm is a mathematical transformation or formula. There are few strong algorithms, and most of them are published as standards or in mathematical papers. The key is a string of binary digits which has of itself no meaning. Modern cryptology assumes that the algorithm is known or can be discovered. It is the key that is kept secret and that changes with every implementation. Decryption may use the same or a different pair of algorithm and key.

The plaintext is often rearranged before being encrypted; this is generally known as *scrambling*. More specifically, *hash functions* reduce a block of data (which can in principle be of any size) to a predetermined length. The original data cannot be reconstructed from the hashed value. When encryption is used to authenticate the other party, the result is usually known as a *digital signature*. Hash functions are often required before a signature is generated; a *digest* of the message (including the most important elements such as the message number, date and time, and main data fields) is constructed and hashed prior to signature.

A *message authentication check* (MAC) is a fixed algorithm for generating a signature on a message. Its purpose is to demonstrate that the message has not been altered between origination and reception.

Algorithms

The design of cryptographic algorithms is a matter for mathematical specialists. The smart-card system designer must, however, be aware of the strengths and weaknesses of the specific algorithms available and be able to decide when each is appropriate. Although cryptology has advanced immeasurably since the pioneering work of Shannon in the late 1940s and early 1950s, cryptanalysis goes hand in hand with encryption, and few algorithms stand the test of time. So the number of algorithms used in practical computer systems, and particularly in smart-card systems, is very small. The main ones are described below.

Symmetric key systems

A symmetric algorithm is one that uses the same key for encryption and decryption.

The most widely used form of encryption in smart cards (and indeed in almost all data security systems) is the data encryption algorithm (DEA), more commonly known as DES. DES was originally a U.S. government product, but it is now a widely used international standard (see the appendix). Sixty-four-bit blocks of data are encrypted and decrypted using a single key, typically 56 bits in length. DES is computationally quite simple and can readily be performed using slow processors (including those in smart cards).

The scheme depends on the secrecy of the key. It is therefore only suitable in two situations: where the keys can be distributed and stored in a dependable and secure way or where the keys are exchanged between two systems that have already authenticated each other and their life does not extend beyond the session or transaction. DES encryption is most commonly used to protect data from eavesdropping during transmission.

Triple-DES encryption is the process of enciphering the original data using the DES algorithm performed three times (normally twice forwards with the same key and once backwards with a different key—see Figure 5.1). This has the effect of tripling the effective length of the key; as we will see later, this is a critical factor in the strength of encryption.

A newer standard symmetric algorithm has been proposed, called the *international data encryption algorithm* (IDEA) [1], but this is not generally available in hardware and has therefore not challenged DES as the algorithm of choice for microcontroller-based applications.

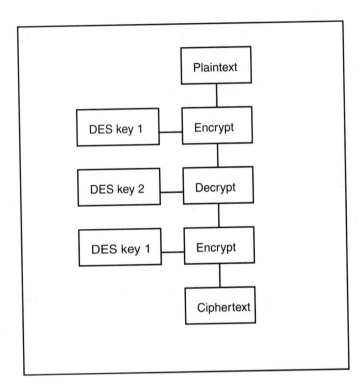

Figure 5.1   Triple DES encryption.

Asymmetric key systems

Asymmetric key systems use a different key for encryption and decryption. Many systems allow one component (the *public key*) to be published, while the other (the *secret key* or *private key*) is retained by its owner. The most common asymmetric system is known as RSA (after the initials of its originators Rivest, Shamir, and Adleman), although there are several other schemes. RSA involves two transformations, both of which require modular exponentiation with very long exponents:

- *Signature* encrypts the plaintext using the secret key;
- *Decryption* performs the equivalent operation on the ciphertext but using the public key. For signature verification, we test whether this result is the same as the original data; if it is,

then the signature was produced using the corresponding secret key.

The basis of the RSA system is the formula

$$X = Y^k (\text{mod } r)$$

where X is the ciphertext, Y is the plaintext, k is the secret key, and r is the product of two carefully selected large prime numbers. For a fuller explanation of the mathematics, see the original paper [2] or any cryptography textbook. This form of arithmetic is very slow on byte-oriented processors, particularly the 8-bit processors typically used in smart cards today. So although RSA will permit both authentication and encryption, it is primarily used in smart-card systems to *authenticate* the originator of a message, to prove that data have not been altered since the signature was generated, and for encrypting further keys.

Other asymmetric key systems include discrete logarithm systems, such as Diffie-Hellman [3], ElGamal [4], and other polynomial and elliptic curve schemes. Many of these schemes are one-way functions, which allow authentication and verification but not decryption.

A particularly elegant and efficient scheme for authenticating a smart card within a system is the zero-knowledge scheme proposed by Guillou and Quisquater [5]. This involves a two-part challenge-response mechanism, as shown in Figure 5.2. In a zero-knowledge test, the entity being authenticated is able to prove its identity (its knowledge of a secret) without giving away the secret. The shadow ID used in the Guillou-Quisquater scheme is much longer than the card ID, while the secret number is a modular function of the card ID. Again, the computation requires modular exponentiation, but exponent sizes can be kept relatively short because the challenge is only used once and there is no plaintext for comparison.

RSA is the basis for the authentication and key exchange functions of Pretty Good Privacy (PGP) and Riordan's Internet Privacy Enhanced Mail (RIPEM), which are systems for protecting the confidentiality of

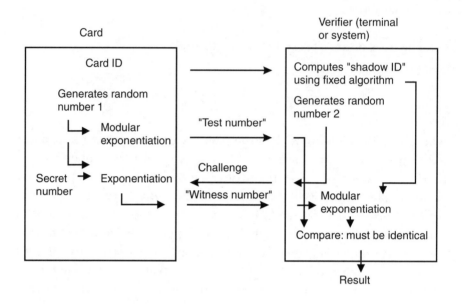

Figure 5.2 Guillou-Quisquater zero-knowledge test.

electronic mail. Both of these use symmetric algorithms, however, for encrypting the body of the text.

Keys

Having established the importance of keys to data security, we must now consider what different types of keys can be used and where they are appropriate.

Secret keys

Symmetric algorithms such as DES use secret keys; the key itself must be transmitted and stored by both parties to the transaction. Because we assume that the algorithm is known, this places great emphasis on the need to transmit and store keys securely. Smart cards are often used to store secret keys. In this case it is important to ensure that the scope of the key is limited: we should always assume that one card may be successfully analyzed, and this must not put the whole system in jeopardy.

Public and private keys

The main advantage of asymmetric key systems is that they allow one key (the private key) to be held very securely by its originator while the other can be published. Public keys can be sent with messages, listed in directories (there is provision for public keys in the ITU X.500 electronic messaging directory scheme), and passed from one person to the next. The mechanism for distributing public keys can be formal (a key distribution center) or informal (the term *web of trust* is used by some authors).

The secrecy of the private key in such a system is paramount; it must be protected by both physical and logical means within the computer in which it is stored. Private keys should never be stored unencrypted, nor in human-readable form. Again, smart cards are used for storing secret keys, but the secret key for the whole system should not normally be stored in a card.

Master keys and derived keys

One way of reducing the number of keys that must be transmitted and stored is to derive the keys each time they are used. Under a derivation key scheme, a single *master key* is used, together with some individual parameters, to calculate a specific derived key, which is then used for encryption. For example, if a single issuer has a large card population, the card number may be encrypted using the master key to produce the derived key for each card (see Figure 5.3).

Another form of derived key is calculated using tokens, which are electronic calculators with special functions. They may use as inputs a challenge issued by the central system, a PIN entered by the user, and the date and time. The token itself contains the algorithm and a master key. Such tokens are often used for secure computer system access.

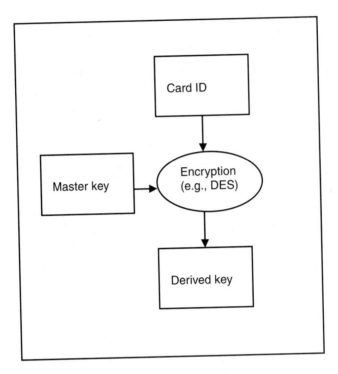

Figure 5.3   Key derivation using a master key.

User and equipment keys

This principle may be extended to produce complete hierarchies of keys. For example, a function (setting up staff IDs or an order file) may only be allowed by one group of users from one specific terminal or group of terminals. Critical parts of this function (e.g., authorizing the production of a staff card or transmitting the orders) may require one individual's personal authority. In this case, we might use a hierarchy such as that shown in Figure 5.4.

Key-encrypting keys

Because the transmission of the key is potentially a weak point in the system, it makes sense to encrypt keys when they are transmitted and to store them in encrypted form. The key-encrypting keys never pass outside the computer system (or card) and can therefore be protected

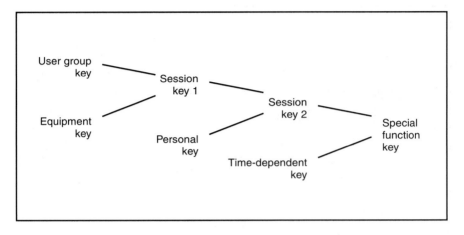

Figure 5.4 Example of a hierarchical key structure.

more easily than the transmission itself. Often a different algorithm is used to exchange keys from that used to encrypt messages.

We use the concept of key "domains" to limit the scope of keys and to protect keys within their domain. Typically a domain will be a computer system that can be protected logically and physically. Keys used within one domain are stored under a local key-encrypting key for that domain; when they are transmitted to another computer system, they are decrypted and reencrypted under a new key, often known as the *zone control key*. On receipt at the other end, they are again translated to encryption under the new system's local key (see Figure 5.5).

Session keys

To limit the time for which keys are valid, a new key is often generated for each session or each transaction. This may be a random number, generated by the terminal following authentication of the card (see Figure 5.6). If the card can perform RSA decryption, then the session key can be transmitted using the RSA key.

The part of the transaction in which the key is transmitted is often quite short in comparison with the rest of the transaction; thus, it can carry a higher overhead than the bulk of the transmission. RSA keys are

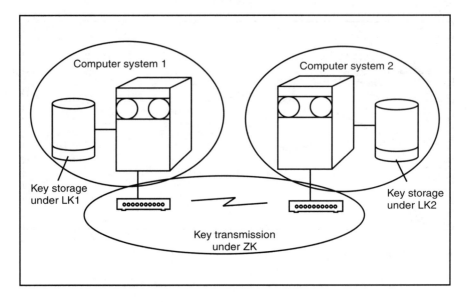

Figure 5.5 Key domains and key encryption.

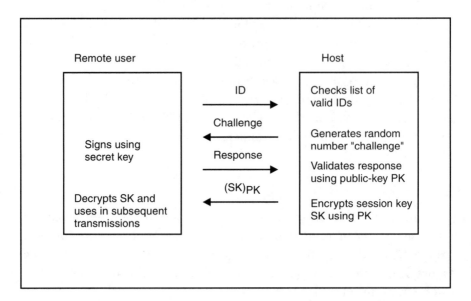

Figure 5.6 Authentication and key exchange dialogue using asymmetric keys.

much longer than DES keys, and the processing time is usually much longer.

A special form of session key is the transaction-key system, used in some electronic payment and electronic data interchange (EDI) systems. Here, a new key is transmitted at the end of each transaction, and this is used for the next transaction.

Selecting an algorithm and key length

No cryptosystem is absolutely secure under all circumstances. There is always a chance that someone may stumble on the algorithm and the key by pure chance. Or, particularly where the algorithm is known, an attacker can try every key combination until one appears to work (a *brute force* attack). This task becomes more time consuming as the key becomes longer, and it should be possible to calculate the time required to do an exhaustive search of the whole keyspace (i.e., all possible key values). For a strong algorithm and suitable key length, this may be several decades of processing using the most powerful supercomputers currently available (although the right answer could be the first key tried!).

The strength of any particular form of encryption depends on the length of the key. This is not only because the brute force attack takes much longer, but also because there is less scope for speculative attacks: feeding in possible combinations of values or trying likely plaintext values.

The criterion for selecting an algorithm and key length is that the cost and effort required to find the key or break the security should be greater than the maximum possible reward.

This criterion is important for card-based systems. With a limited range of algorithms and key lengths available, the other variable within the system designer's control is the maximum reward. In other words, the system must limit the damage that could be done by breaking a single code, so that it is not worth an attacker's time trying to break it. The main ways in which this is done are to:

- Limit the scope of a single key (to one card or one type of transaction);

- Limit the time for which a key is valid. Many keys are only used for a single message or transaction; it is easier to protect a key that is only used to generate other keys than a key used for encryption;
- Include sensitive monitoring functions in the system, so that any irregularity is quickly detected and the cards or parts of the system affected can be isolated.

When symmetric encryption is being used, therefore, it is best to generate keys in a hierarchy, so that only the lowest levels are liable to be compromised. The DES standard (and hence all generally available hardware) uses 56-bit keys. A combination of faster computers and weaknesses discovered since this algorithm was published have resulted in successful breaches of DES, at least in the laboratory. Where a DES key is used repeatedly, therefore, the triple-DES technique (using two keys) should be used. The alternative is to generate a new key for each message or transaction. Banking message encryption has used triple-DES for many years.

With public-key systems, the length of key required depends entirely how the key will be used. In general, the scope for discovering the private key is usually much more limited than for a symmetric key, because it should never be used directly except within the confines of the key owner's own computer. Only I may use my key. However, the private key could be open to a brute force or analytical attack if I have used that key to authenticate (by signing) many messages or other data and the attacker gains access to examples of the signed data and the original.

Public-key systems based on the RSA principle depend on a number that is the product of two large prime numbers. If this number can be factored into its components, the key can readily be broken. So the key should be as long as practical, subject to the maximum acceptable transaction time in the application concerned. For access to a secure building or computer, for example, a transaction time of several seconds might be acceptable, but when boarding a bus anything over 150 ms results in a large number of failed transactions and impatient passengers.

For comparison, an RSA key of 64 bits was broken in 1995 using three linked massively parallel computers over four days; a 40-bit key in 1996 required three-and-a-half hours with a network of 250 workstations. The time required for factorization increases exponentially with the length of the key (a 65-bit key should take twice as long as a 64-bit key), so unless new factorization methods are discovered (this is an important reservation) the keys used today, typically 512 or 1024 bits, have a substantial margin of safety.

It remains important, however, to ensure that the key is not compromised by other means. This is the purpose of key management.

Key management

As with the locks on a building, users of encryption systems must make sure that their keys are created and stored securely, and that they are only available to properly authorized personnel.

Key generation

Symmetric keys should normally be generated completely randomly. Many so-called random-number generators generate not a random number, but a fixed sequence. This characteristic can be useful in some situations (such as the *code guns* used in radio systems), but if the source and starting point can be determined, then the entire system is compromised. True random-number generators typically digitize some of the *white noise* generated by a free-input amplifier.

Asymmetric key generation is a more complex process, as it requires the use of large prime numbers and not all combinations are suitable. The process must still be seeded using a random input, however, or the same problems will ensue. In addition, it has been found that some combinations of keys are easily broken, and the key set produced must be tested for these conditions before it is used. These processes are normally carried out automatically within the key-generation system that forms part of every public key encryption package.

The process of key generation must be carefully controlled. In many organizations, it takes the form of a small ceremony: Several people are involved and each checks the operations of one other. No one person has access to the whole sequence.

Key transmission

The problems of key transmission have already been mentioned and some of the solutions described. Regular transmission of session keys is handled by encrypting them under a key-encrypting key during transmission. Where a master key must be loaded into several pieces of equipment, it is now normal to connect them to a key server and to load the keys directly using the same technique.

In some systems, it is still necessary to transfer keys manually. In this case it is usual to divide the key into three or more parts; each part is transferred to paper (or committed to memory) by one person and loaded by the same person at the other end. Whether the process is carried out manually or automatically, a high level of formal control and discipline is required during this process.

Key transmission is less of a problem with asymmetric key systems because only the public key is normally transmitted from one system to another. Even then, it is desirable to use a message authentication check (MAC) or some other form of integrity check to ensure that the key is not altered during transmission.

Key indexes

By now it will be clear that there is an underlying assumption that keys may be discovered, even if we do everything we can to avoid this. We must have a plan for detecting compromise and know how we will react if this happens. We have also said that one way to limit the damage caused by a breach of security is to limit the time for which any key is valid. Both of these requirements point to the need to set up more than one set of keys and to have a mechanism for moving from one set to the next.

This is done by generating several sets of keys, each of which is associated with an index. When the key is stored, the index remains with it. Cards and other remote systems may be loaded with three or more keys, and they advance to the next key when instructed to do so by the host system (for offline systems such as smart cards, this will take place at the beginning of the next transaction). The new key set need never be transmitted.

There is a particularly high danger of key compromise during the testing phase of any system, so it is wise to advance the index after tests are completed. And when any major improvements to security have been made, the benefit of the improvement may be lost if the existing keys remain in use—so this is another occasion on which a new set of keys should be used.

Certification authority

A *certification authority* (CA) is a trusted computer system that is able to testify to the identity and authenticity of one or more of the parties in a transaction. Often this is the owner of the scheme. Certification can be an online or an offline process; either the entity presents its ID together with a certificate from the CA or it presents its ID and the other party seeks certification online from the CA. The process may be mutual: both parties may require authentication of the other.

Where authentication is offline, there should usually be a mechanism for registering invalid certificates (where an organization no longer belongs to the certified group). This may be done by giving certificates a limited life or by transmitting lists of lapsed certificates to those still subscribing.

The CA may also act as a central repository or distribution center for public keys. To guard against incorrect public keys being published or errors being introduced, public keys may be signed by the CA, using its secret key. They can then be decrypted using its public key, which is assumed to be available to all.

Computational requirements

All cryptology requires mathematical manipulation of blocks of data. Even general-purpose computers can be slow at these computationally intensive tasks, and various functions can help in this process.

DES is available directly in hardware: With its fixed length input and key size, it is well suited to large-scale integration, either on its own or in conjunction with other cryptographic functions. Several manufacturers offer this function.

For RSA and similar functions, there are two possible approaches: to provide a fast arithmetic unit (which provides rapid execution of multiple-length arithmetic functions), or to provide special functions adapted to the solution of expected encryption problems.

Most specialist cryptographic hardware (including cryptographic microprocessors used in smart cards) now adopt the second approach. Although it is less flexible, the performance improvement over the general purpose coprocessor is very substantial. Cryptoprocessors are able to incorporate special functions such as the Chinese Remainder Theorem, which provides a fast route to the modular exponentiation required by RSA.

The performance of smart cards when performing cryptographic functions will be discussed in more detail in Chapter 8, but chips are now available (for use in terminals and host systems) that will perform DES calculations in under 1 ms (i.e., in real time at 64 Kbps) and 1,024-bit RSA in under 500 ms.

DES is also widely available in software form, including several sources that can be downloaded from the Internet. Although several of the DES standards exclude software implementations in general-purpose computers, these are nonetheless widely used.

Cryptography export controls

Many governments treat encryption products as weapons and seek to control their use and particularly their export. The U.S. and French governments have particularly stringent regulations. The data communications industry has fought against these restrictions, arguing that their effect is to give criminals access to stronger security than law-abiding citizens. The wide availability of many algorithms in published—and even in software—form means that the controls are not completely effective, although they do constrain the international flow of some implementations and restrict the length of keys that may be used in other cases. Most commercially available products use licensed forms of the relevant standards and are not subject to further controls.

From a practical point of view, much of the most advanced work on cryptography is carried out in government laboratories such as the National Security Agency (NSA) in the United States and the Computer and Electronics Security Group (CESG) in the United Kingdom. They are able to control what is published.

The U.S. government sought in the early 1990s to impose a hardware-based encryption scheme known as Clipper. This was fiercely resisted by the cryptology community, and Clipper is now only mandated in a small number of cases.

A further proposal was in the form of the key-escrow arrangements, under which any user of "strong" encryption products would be required to lodge all keys with the appropriate government departments. This has so far proved impractical to implement, and it raises a wide range of opposition from civil liberties and consumer protection groups.

Summary

For the smart-card system developer, a few key facts can be extracted:

- There is an encryption method available to handle almost any security requirement. The normal systems designer should make use of commercially available products rather than trying to design a unique system; such products will be licensed and meet the relevant requirements for encryption usage and export.

- The requirements for card authentication, data confidentiality, and message integrity should normally be considered separately. Authentication can be performed using a symmetric algorithm such as DES or an asymmetric (public-key) system such as RSA. Data stored within the card can be considered confidential provided that it is protected by a suitable chip and access-control mechanism. However, if it passes outside the card, it may be encrypted, probably using DES or another symmetric algorithm. Message integrity should be ensured by using a standard message authentication check.

- DES is computationally simpler than RSA; it is available in hardware and in several smart-card chips. However, it requires a secret key to be shared between the card and the decrypting device; thus, it is not generally suitable for card authentication in open systems.

- RSA requires more computation, and is only available at reasonable speed in the fastest smart-card chips; however, the ability to publish one key openly, while the other key remains secret inside the card or host system, is the reason it is more commonly used in open systems. Simpler chips can store a certificate created with the secret key, which can be checked by a terminal.

- The strength of any encryption system depends on the length of the keys used; the keys used in typical applications will be discussed further in Part 3.

References

[1] Lai, X., and J. L. Massey, "A proposal for a new block encryption standard," Advances in Cryptology, *Proc. Eurocrypt 1990,* Springer-Verlag, 1990.

[2] Rivest, R., A. Shamir, and L. Adleman, "A method for obtaining digital signatures and public key cryptosystems," *Comm ACM,* Vol. 21, No. 2, 1978.

[3] Diffie, W, and M. E. Hellman, "New Directions in Cryptography," *IEEE Transactions on Information Theory,* Vol. 22, No. 6, 1976.

[4] ElGamal, T., "A Public Key Cryptosystem and a Signature Scheme based on Discrete Logarithms," *IEEE Transactions on Information Theory,* Vol. 31, No. 4, 1985.

[5] Guillou, Quisquater, *Zero Knowledge Identification Proof for Smart Cards,* 1988.

6

Passwords and Biometrics

THIS CHAPTER IS concerned with the many situations in which we need to be able to identify or verify the identity of a person. We are mostly concerned with checks that can be carried out automatically by a system, although the system may also be used to provide the evidence for another person to check.

Personal identification types

Passwords, tokens, and biometrics

Virtually all identification methods involve something you *know* (a password or PIN number), something you *have* (a card or other token), or something you *are* (a physical or behavioral characteristic). The last group is known as biometric checking.

Each has its advantages and drawbacks. Passwords can be learned by subterfuge, guessed, or given away, but they do allow the user to delegate authority. (Elderly or disabled people often ask others to draw money using their ATM cards, and it is sometimes useful to be able to ask someone else to look up some data on our computers when we are out of the office.) Many people simply forget passwords, particularly if they use them infrequently. Tokens can be lost or stolen, or again they can be lent or transferred to someone else deliberately. Physical characteristics are inflexible; they are often difficult to transmit down telephone lines, for example. Security system designers must ask themselves whether users should have the ability to delegate or transfer their rights to someone else, as this will affect the choice of identification type.

Some identification methods can only be used in a face-to-face mode, when both the person seeking acceptance and the person checking the identity are present. Others may be used when one or other is absent: This would be important for cash withdrawals at an ATM, say, or for remote access to a computer system. In this chapter we are mostly concerned with methods that will allow an automatic check, without any human involvement, as this is the form of identification normally addressed by smart-card systems.

Identification methods can be used in combination: a card and password or card and biometric would be common. The combination may be varied according to the requirements: For example, we might use a card only for entry to the building, a card and a PIN for the computer room, but a card and a fingerprint for a funds-transfer operation on the computer system.

Where several characteristics are known, the combination can be varied dynamically: the user presenting a card might be asked for additional identification depending on the transaction or the results of previous checks. This is particularly useful in avoiding false rejects (see the next section).

Behavioral and physiometric

Biometric checks are divided into:

- *Behavioral techniques* that measure they way we do something such as signing our name or speaking a phrase;
- *Physiometrics* that measure a physical characteristic such as a fingerprint or the shape of a hand.

Behavior changes with time and mood. Behavioral measurement techniques work best when they are used regularly, so that these changes, and each individual's level of variability, can be taken into account. The models for behavioral measurement have to take into account these ranges. Physiometric measurements, on the other hand, require larger measuring devices and more complex software. They have to match the position of the hand, for example, with the template (it will never be identical).

There are psychological differences between the two groups as well: With a behavioral method, users typically feel that they are in control. With a physiometric system, on the other hand, users sometimes feel threatened by something that denies them entry or use based on a measurement they cannot control.

Requirements

Recognition versus verification

We must first distinguish between systems designed to recognize people (is this a person I recognize and, if so, whom?) and those that must only confirm a person's identity (is this the person he or she claims to be?). This second task is much easier and the verification parameters can be set on an individual basis. This is also the normal case for smart-card systems, whether the reference pattern (usually called the *template*) is held on the card or on a central system.

Performance

The most important performance characteristics of a biometric are its ability to reject impostors while accepting all valid users. These are usually measured by the false-accept rate (FAR)—the percentage of impostors accepted by the system—and the false-reject rate (FRR)—the percentage of valid users rejected. Most systems can be tuned to provide either sensitive detection (low FAR but high FRR) or coarse detection (low FRR but high FAR)—see Figure 6.1. The critical measurement is known as the *crossover rate*—this is the level at which the FAR and FRR are identical. Most commercial biometrics today have crossover rates

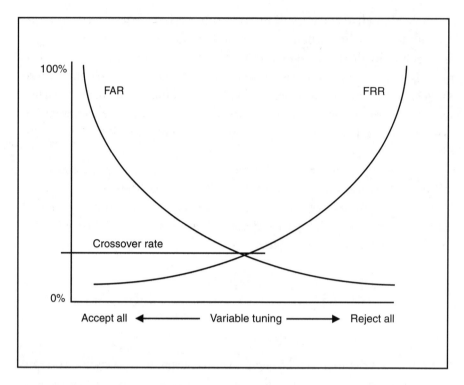

Figure 6.1   False acceptance vs. false rejection in a biometric.

below 0.2%, and some are below 0.1%. The rate improves with frequency of usage, as users become accustomed to the system and the
system becomes better tuned to the level of variability expected.

The balance between these two extremes depends on the application:
will it be used by the public or by trained employees? Are there customer-
service issues or do users have little choice? Typically, banks seeking an
ATM biometric will opt for a higher false-accept rate to avoid offending
valid customers, whereas a welfare agency checking applicants will demand a very low false-accept rate, even if this means extra time to check
those rejected.

Procedures

A biometric test normally involves three stages: enrollment, use, and
update.

Users are enrolled in the system by an initial measurement. This will usually be performed three or more times to check for consistency; the procedure often takes longer than the measurement for testing purposes.

In normal use, the user must be guided to help provide consistent results; it is no use, for example, enrolling a signature on a large plate placed at a convenient angle and then expecting similar results from a field unit that is much smaller and located under a privacy shield. Once the measurement is captured, it is compared with the template; this is where it is important to allow appropriate levels of tolerance, particular for behavioral measurements.

Most biometric systems, particularly those that use a behavioral characteristic, must have a provision for updating the template. In the case of signature and voice recognition, this is usually an adaptive function, performed by the software every time the measurement is checked. For slower moving characteristics, the system may monitor the percentage fit or the number of times a person is rejected and seek a reenrolment when necessary. A transaction log is often a useful feature and can easily be incorporated in a smart-card-based system.

Components

A biometric system will include:

- *A measurement device,* which forms the user interface. Ease of use is another important factor for biometrics: the device must be instinctive and leave little room for error. It must be suitable for use by a wide range of people, including those with disabilities.

- *Operating software,* including the mathematical algorithms that will check the measurement against the template. The newest algorithms depend less on statistical modeling, and more on dynamic programming, neural networks, and fuzzy logic. This increases their flexibility; they are less likely to reject someone because of a patch of dirt, for example, if the rest of the pattern fits closely.

- *External hardware and systems:* The usability, reliability, and cost of the system will often depend at least as much on these external systems as on the measurement device. Some systems (such as fingerprint checking) are intrinsically well suited to use in distributed systems, while others (such as voice recognition) are more appropriate for centralized systems.

Training and user instructions also form an important part of a successful biometric implementation. Where the system must be used by the public, the emphasis will be on the user instructions and design of the measurement station; with staff and other closed user groups, more benefit will be obtained from training.

Most biometric techniques cost around $1,000–$3,000 per station today. This sum is falling with each new generation of equipment, and devices will become available shortly for under $200. At this level, it becomes practical to fit them to ATMs and specialized access control devices. A further drop in cost, to below $100, will probably be required before it would be viable to use biometrics in a retail, vending, or regular access-control environment.

Passwords and PINs

The best known and most common form of personal identification is the password. Passwords on their own are not well suited to high-security applications: people forget them, write them down, or use easily guessed words, and they can be monitored and replayed. A very high proportion of successful computer system hacks involve the discovery of passwords by one means or another.

Much better results are available when passwords are combined with some form of token or challenge-response generator, so that the password is not input directly to the system. This removes one of the biggest advantages of the password—its cheapness—and does not overcome the limitations of users' memories.

A PIN is a simple form of password, consisting usually of four to six digits, which can be used with a numeric keypad rather than a full keyboard. They are familiar to most of us from their use in conjunction with bank cards. Many people cannot remember even a four-digit PIN, particularly if they have several cards with different PINs; thus, the ability to change a PIN is important. Many users write their PINs down and keep them close to the card, where they can find them when required. All of these factors reduce the value of the PIN as a safety mechanism, but they can be overcome by suitable education of users. Nevertheless, PINs should only be regarded as a secondary identity check; the card is the primary identification.

Passwords and PINs share the advantage that they are either right or wrong: There is no latitude for error and hence the software required to check them is very simple. The opposite is true for most biometric checks.

Behavioral

Signature verification

Automatic signature checks are a natural extension of a familiar process: no cultural change is required. Whereas the human operator checks the final shape of the signature, however, most automatic forms of signature verification place much more emphasis on the dynamics of the signature process: the relative speed with which lines are drawn and the pressure used. This enables them to compare measurements between signatures even where the environment is quite different and defeats most attempts at forgery.

There are many proprietary methods for measuring and storing signature dynamics, several of which have been patented. Each device measures different aspects of the signature and attempts to differentiate between stable aspects and those that vary from one signature to the next. The measurements are made using a digitizing tablet or a special writing surface.

As with all behavioral characteristics, signatures are subject to the mood of the user, the environment, the pen and paper, and so on. Some people's signatures are very consistent, where others vary greatly. By careful system tuning, very good results can now be obtained for regular users, but there are often a few "problem users" who are consistently rejected.

The signature template is typically 1 kb; this makes the technique suitable for online use or with a smart card. A large-scale pilot was very successful in reducing benefits-claim fraud in a U.K. unemployment office.

A subsidiary advantage of most signature-verification systems is that they capture a record of the signature as proof of the transaction: this allows *truncation* of paper systems such as credit card vouchers, since the electronic record of the signature removes the need for the paper voucher.

Keystroke dynamics

The way a person types at the keyboard is almost as characteristic as the way they sign their name; skilled typists in particular can be recognized almost instantly from their typing patterns. This offers a quick way to identify computer system users, without requiring any additional input device or enrollment procedure. Current implementations are limited to the laboratory, due to the problems of variations between keyboards and system software delays. On the other hand, the low overhead and transparent operation make this a very attractive technique for applications such protecting a small number of supervisor-status users on a computer system. Although the applications of such a biometric are clearly limited, this is potentially a very useful additional tool in the armory of computer access control techniques.

Voice recognition

Voice recognition systems are the biometric most readily accepted by customers, but unfortunately they have not yet reached the levels of performance required in most commercial environments.

Using voice recognition allows multilevel checking: the system can check what is said as well as how it is said. In some environments it is very low cost: a single digital signal processor (DSP) can process for many microphones. Voice recognition is a branch of speech processing technology, which has many wider applications in other fields, notably digital telephony and videoconferencing. Use is made of spectral analysis and specialized algorithms that are being developed for high-density compression and accurate representation of speech.

High levels of false rejection are a problem with voice recognition systems, often due to colds or unusual background noise conditions, but one neural network system has achieved a crossover rate of better than 0.1%. Impersonation is less of a problem than might be thought because the aspects of the voice measured by the systems are not the same as those a human listener is likely to notice, whereas an impersonator concentrates on the human characteristics. One of the most important factors in voice recognition is the size and shape of the "voice box" in the throat.

Voice recognition can be used over telephone lines and has been used by Sprint in the United States since 1993. Although it could be used with smart cards, we are not aware of any commercial schemes that use this combination yet.

Physiometric

Finger/thumbprint

This is the oldest form of remote identification test; although it was previously associated with criminality, research in many countries is now showing a level of acceptance that would allow its use in public schemes.

Systems can capture the minutiae of the fingerprint (points such as intersections or ridge endings) or the whole image. Minutiae are usually captured in the form of x and y coordinates and a direction; if these three variables are used, the resolution can be quite coarse. This leads to templates of around 100 bytes, compared with 500 to 1,500 bytes for a full image of the fingerprint. A differential method has also been developed; this produces interference fringes which can be encoded in an even smaller template (around 60 bytes). Crossover rates of less than 0.05% can be achieved.

There are some problems with consistency in public schemes; workers in some trades, as well as heavy smokers, often have very faint prints which are difficult to analyze repeatably. Nonetheless, there have been many successful and long-running schemes using fingerprints, including a high-profile trial at the World Expo in Seville in 1994.

Current readers cost between $500 and $1,500. The next generation will be able to use specially designed single-chip sensors able to capture variations in capacitance across an array the size of a finger, and these should reduce the cost to under $150. With sufficient volume, it should be possible to manufacture readers for under $50.

Hand geometry

Hand geometry scores highly for ease of use because the hand is a large feature and can be placed very consistently using a guide system. Hand-geometry measurement devices of different sorts have been in use since the early 1970s, making this the oldest form of biometric. The hand is guided into the correct position for scanning using a set of pegs, and the image is captured using a video camera. The reference template can be made very small (one commercial product uses only 10 bytes).

Although day-to-day variations such as dirt and cuts do not affect its performance, this measurement is affected by injury and aging, and reenrollment may be required from time to time.

High-profile applications have included access to the athletes' village at the 1996 Olympic Games in Atlanta, frequent traveler controls at several U.S. airports, and for voting in the Colombian legislature. Hand geometry is also used for many other access-control applications.

Retina scan

Retina scanners measure the characteristic blood vessel patterns on the retina (the back of the eye) using a low-power infrared laser and camera. This technique involves placing the eye close to the camera to obtain a focused image. It has been used for several years to control access to a number of U.S. military establishments. Templates are very small (35 bytes in the most common commercial system) and the crossover false-accept/false-reject rate is very low.

Recent medical research has shown, however, that retina characteristics are not as stable as was once thought: they are affected by disease, including diseases of which the owner may not be aware. Many people are worried at the thought of placing their eye in close contact with a light source and the damage this might cause. As a result, this technique has largely given way to the less invasive iris scan.

Iris scan

Iris scanners measure the flecks in the iris of the eye. The technique has a very high level of discrimination as a result of the large number of features and is found to be very stable with time.

The user has to look at a camera from a distance of 30 cm or more for a few seconds. The system accommodates spectacle and contact lens users without difficulty, although the sensor must be mounted or adjusted in such a way that it is suitable for users of different heights, including those in wheelchairs.

Iris scanning is a newer technology than most of those described; it is used in several U.S. prisons, where the exceptionally low FAR is a major advantage.

Others

Almost any part of the human anatomy could in theory be used in a biometric. Commercial schemes have concentrated on those that are easily measured and are likely to be most socially acceptable.

The story of the Indian family who collected father's pension for years after his death by using his finger may be apocryphal. Nevertheless, it is sometimes necessary to check that the measurement is being taken from a live person, rather than a copy. Some devices includes checks for this (such as movements in the flecks of the iris or subcutaneous structures). One method that uses a living characteristic directly is vein scanning: This measures the position of the veins from the flow of warm blood.

Facial recognition—the technique humans use most—is a viable biometric, particularly where cameras are already in use and where a very coarse check is all that is required. Specific features or *hot spots* are measured and used to create the template.

The ultimate biometric is DNA analysis. However, it can safely be said that it will be many years before this technique is used for checking identity in the average store or when boarding a bus!

Biometrics and cards

Some identification techniques, notably passwords and PINs, are exceptionally well suited to implementation in a distributed system. They have minimal storage and processing requirements, but as we have seen they are not regarded as secure on their own.

Many systems make use of a card in conjunction with another technique. The minimum requirement is to provide a reference or identification number (this is who I am claiming to be); the template or reference password is held on a central system. However, in other applications, there are advantages in making the template portable. This allows the check to be carried out by offline terminals, and a cardholder's ability to carry his or her own template can make the system seem less like Big Brother is in control.

The storage requirement of the template is thus a limiting factor; as we have seen, most templates require between 40 and 1,500 bytes. The smaller templates can be stored on a magnetic-stripe card or barcode, albeit with the limitations of those technologies mentioned in Chapter 4. For larger templates, two-dimensional barcodes or holograms may be used, but the most satisfactory answer in almost every case where secure storage is required is to use a smart card.

In the next three chapters, we will explore the characteristics of the smart card that make it particularly suitable for storage and checking of biometrics and passwords.

7

Smart Card Types and Characteristics

Smart cards fall into several categories. Within each category there is a range of chip capacities, power and speeds. In this chapter we consider the broad product groups, and in Chapter 8 we look at the quantitative factors that distinguish smart-card products. We are primarily concerned here with differences that affect security; there are several other publications, some of which are listed in the bibliography, that describe other smart-card characteristics.

Smart cards can be divided into groups by:

- *Function:* The fundamental difference is that between memory cards and microprocessor cards, and there are further specialized functions within each group.

- *Access mechanism:* Contact or contactless.

- *Physical characteristics:* Shape and size.

Memory cards

Unprotected

The simplest smart cards are not smart at all: They contain memory circuits that are directly accessible through the contacts using a synchronous protocol. Some early telephone cards used this scheme, but they are not widely used today.

Protected

Most current memory cards contain, as a minimum, a protected area. The memory in the card—usually *electrically erasable programmable read-only memory* (EEPROM or E^2PROM) is divided into two or more areas.

One part is accessible immediately when the card is powered up—in fact, it will often be transmitted to the terminal or host system as a part of the card's response when it is powered up; this is known as the *answer to reset* (ATR). But the other data are only accessible after the card has received a security code, either from the cardholder or from the software application, which matches a code held secretly in the card. Often there will be several such protected areas: one part may be accessible after the cardholder has given a code, another is unlocked by a program controlled by the card issuer, and a third part is only accessible to the manufacturer. Sometimes the codes themselves are held in a further area, which is only accessible once, at the time when the card is issued or personalized. Once the codes are written, a fuse is blown to prevent them from being altered.

Protected memory cards are available with memories from under 100 bits to several kilobits. Memory sizes are most often quoted in bits, as these units are most commonly used in counting applications, such as telephone cards and vending. They can also be used for simple access control with a single numeric key or PIN.

Basic protected memory cards are inexpensive ($1 to $5, even in hundreds) and can be bought and programmed by a fraudster using almost any PC-connected smart-card reader. Assuming that he or she has access to the memory mapping and that there is no further encryption in the system, a counterfeit card which will appear to the system like a real card can be produced. The data on a valid card are protected by the PIN and other security codes, but these are subject to the limitations of all single-code schemes: fraudsters can often find ways of obtaining legitimate PINs through carelessness or deception.

Secure logic

A further group of memory cards not only controls access to memory, but also restricts the ways in which the external application can write to or read from the memory. For example, prepaid telephone cards can contain logic to prevent the value from being increased. They may also contain special logic that enables the application to authenticate the card cryptographically by holding a *hashed* form of the card data (see Chapter 5) which alters as the value is decremented.

Secure logic chips are also likely to include several of the special security features described in the next chapter, including irreversible programming, transport codes, and logical address scrambling. Many also contain specific provision, known as *antitearing,* to protect the data against corruption if the card is removed prematurely or if power is removed from the card before it has completed the transaction.

Most modern telephone cards, public-transportation cards, and pre-payment cards follow this pattern. They are considerably more secure than the basic protected memory cards, and if the correct issuing procedures are followed (see Chapter 10) the risk of counterfeit cards should be virtually eliminated. It is also important that the authentication mechanisms built into the cards are used. Otherwise it could still be possible to design a card or system that would respond correctly, even if it were not a true copy (for example, it might always give a "true" output, regardless of the input). Designers of smart-card security schemes using this type of card must be able to implement a basic public key encryption program, as described in Chapter 5 and in several of the references in the bibliography.

Microprocessor cards

Development

For maximum security and true portability of data, the card must incorporate a microprocessor. With a microprocessor card, the data are never directly available to the external application: They must always pass through the microprocessor, which can carry on a dialogue with the application.

In the first smart cards, the microprocessor and memory were separate. One of the early Bull patents involved incorporating the two into a single chip, and nowadays several IC manufacturers are able to produce specially designed chips that meet all of the requirements of security, power, and memory capacity within the required thickness of chip.

Conventional

Most smart card microcomputer chips adopt a conventional *von Neumann* architecture, as shown in Figure 7.1. As well as the processor, memory, and power control, there are special circuits for the security and for communication with the outside world. Within the card, data are passed through a bus, under control of the security logic.

The partitioning of the microcomputer's memory, and the logic that governs access to the partitions, is central to the security of a micro-

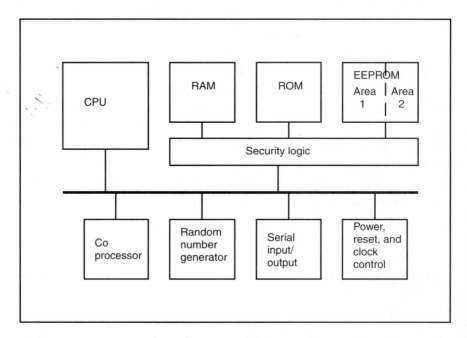

Figure 7.1 Smart card microcomputer architecture.

processor card (although, as we will see in Chapter 8, most smart-card micros also make use of other techniques to prevent unauthorized reading, counterfeiting, or masquerade). All access to memory therefore passes through the security logic. The most important areas of memory—containing the card's own secret keys—are not accessible from outside the card at all.

Many smart-card micros also have an arithmetic coprocessor, which helps particularly with cryptographic functions. Some also have a random number generator, which is used when two-way authentication is required.

There is a single input-output interface, which may take various forms. And every smart card has some form of power-control and clock-control circuitry. This varies in complexity quite considerably; where the card issuer has little or no control over the environment or readers in which the card will be used, the card must be prepared for a wide range of clock speeds and voltages.

From a security point of view, virtually any smart-card microprocessor can be regarded as impossible to counterfeit. Legitimate cards or chips can be bought, however, or they can be stolen at any stage in the manufacturing or personalization cycle. The procedures for card manufacture, personalization, and issuing must ensure that cards are never in a state where keys or confidential data could be extracted or where stolen cards could be programmed with apparently valid keys.

State change

Smart cards using the conventional architecture described earlier are reaching current limits of speed and memory capacity. These limits are set by the power generated by each transistor junction, which must be dissipated through the chip, and by the maximum physical size the chip can be before it becomes too fragile. These limits are gradually being extended, but a solution that meets the requirements of some applications is already available.

Instead of using a conventional microprocessor, which executes microcode through a single set of registers, *programmable gate arrays* (PGAs) have a large number of transistor junctions wired to perform logic according to the *program*. The output depends only on the state of the inputs (which may include memory devices); thus, they are called *state* or *state change* devices rather than microprocessors. PGAs, which occupy a smaller chip area, have been programmed to emulate popular microprocessors and will often be faster at given tasks.

Field programmable gate arrays (FPGAs) can also have user programs added to the logic, and are therefore almost as versatile as the original devices.

PGAs and other-state change devices have the added advantage that they are not in general subject to the patents and license conditions of the microprocessor manufacturers (although care must still be taken in this area). In the case of contactless cards, there may be a further advantage: having neither a microcontroller nor contacts, state-change devices do not appear to be covered by the general patents covering those aspects of smart-card technology. This allows manufacturers and system designers to create smart-card systems free from license requirements.

Against these arguments, state-change devices are easier to analyze and emulate than microcontrollers with their programs. They are typically made in smaller volumes and do not benefit from the special techniques and controls that are applied to the higher end smart-card microcontrollers.

Cryptographic

A further group of smart cards is used for applications that require the card itself to perform specialized cryptographic functions. As we saw in Chapter 5, many of these functions require arithmetic on very long operands, for which most general-purpose processors are too slow.

Such cards therefore incorporate arithmetic coprocessors, incorporating special functions which not only allow multiple word arithmetic (up to 1,028 bits in some cases) but also special algorithms for solving common cryptographic problems.

Crypto cards also incorporate the full range of security devices described in Chapter 8, such as memory-address scrambling, auto-detection of hacking attempts by multiple PIN tries, power-circuit manipulation, or even electron microscopy.

They represent the pinnacle of commercial microcomputer and system security technology in the mid 1990s. Although in most cases these cards are not provably secure in a mathematical sense (we cannot demonstrate that no modes of failure that would compromise security exist), they do appear to defeat in a practical sense every mode of attack currently envisaged. Several teams of cryptographers, physicists, and engineers have set out to break them using massive computing resources, but so far all of the modes considered involve very low probabilities of success.

Contact and contactless

Contact cards

Until now, we have generally been considering cards that will be used as replacements for magnetic-stripe cards, principally in financial and access-control applications. Most of these cards will make their interface with the outside world through a set of six to eight contacts, as defined in ISO 7816 part 1 or in the European Telecommunications Standards Institute (ETSI) or earlier French standards. All early smart cards, in Europe at least, were of this type and met one of these standards.

It is recognized that the contacts themselves are a potential point of weakness in a smart-card system:

- The contacts can become worn through excessive use or damaged by a defective reader or in a pocket;
- The leads from the microcircuit to the contacts are of necessity very thin and can break or become detached when the card is bent or otherwise stressed;
- They represent an obvious starting point for any attack;
- The contact set in the reader is a mechanical device, which can break or be damaged either accidentally or maliciously. Chewing gum and cyanoacrylate adhesives are both popular tools for vandals.

Contactless cards

To overcome these problems, manufacturers have developed contactless cards, in which the microcircuit is fully sealed inside the card and communicates with the outside world through an antenna wound also into the thickness of the card. Power may be provided by a battery, but the preferred solution (adopted by most current designs) is for power to be collected by an antenna, which may be the same as or separate from the communications antenna. The antennas may be quite small (less than 1 cm diameter) or wrapped around the outside of the card, but they can collect sufficient power from the electromagnetic field to power the IC as well as providing the communications path.

The systems can be tuned so that the reading distance is large (up to 50 cm) or small (the card must be placed in contact with the reader). However, this sensitivity is mostly a function of the reader rather than the card, so cardholders could, in theory at least, be justified in fearing that transactions could take place on their cards without their knowing it.

The transportation industry, particularly public transportation, is a major potential user of smart cards. Applications in this industry depend in nearly every case on the use of contactless cards, which offer a much higher transaction rate per terminal because the card does not have to be inserted and removed. Contactless cards and the associated readers are also more reliable, require less maintenance, and should have a longer life than contact cards.

Against this range several arguments, most of which are commercial rather than technical:

- The speed of communication for a contactless card depends on the power of the reader/transmitter and on the reading distance. Transaction times must be kept short, as the card may move out of range very quickly. Contactless cards are therefore unsuitable for applications where significant amounts of data must be exchanged between card and reader.

- It is difficult to make a contactless card as thin as a contact card, but because it will rarely be placed into a reader, this dimension is unlikely to be critical. However, it does mean that manufacturers often set their own standards or produce completely proprietary products.

- Although in principle any smart-card microcircuit could be used in a contactless card, in practice the number of manufacturers is small and the range of card types available is limited. It does not, for example, include any highly secure crypto chips. Large volume purchasers prefer a product that is available from several sources.

- The banking industry in particular has reservations about any card that can carry out a transaction without being inserted into a terminal.

- Contactless cards are currently much more expensive than corresponding contact cards. This is largely because fewer cards have been made; as the volume builds the price differential will shrink but not disappear.

For all of these reasons, contact cards are still preferred for the vast majority of applications. Nevertheless, the market for smart cards in public transportation will not grow until contactless reading is available from a wide range of sources and in a variety of card types. Transportation operators are seen as a very important potential market by most smart-card manufacturers; thus, contactless cards will undoubtedly become much more common over the next few years.

Combi cards

For consumers, the most frequently quoted advantage of the smart card is its ability to handle multiple applications on a single card, thus reducing the number of cards in people's wallets and the confusion this can cause. For some applications, the ability to share cards is also necessary because of the high cost of cards and the need for a network of readers. But if one group of applications requires contact cards and another requires contactless, the two may be unable to share a card.

There are two solutions to this problem; both overcome some of these problems but require that we can read the same card using either contacts or contactless reading.

Pouches that will accept a contact card and effectively convert it into a contactless card are available. The pouch makes contact with the card's contacts, but it has its own power and communications circuitry driven by an antenna. The combination has the speed limitation of the contactless card, but any contact card can be used. This would, for example, enable a bank card also to contain a public transportation application.

While pouches are certainly convenient, they offer from a security point of view the least satisfactory answer: The contact point of weakness still exists, whereas now much more data is being passed between card and reader by radio (which can be intercepted). Current designs do not allow the card to tell whether it is being addressed by radio or by contact, and a radio reader could therefore carry out a transaction that should have been restricted to the contact mode. Pouches should be seen as a temporary solution suitable for lower security applications.

True combination cards have both contacts and an antenna. Some designs have two independent microcircuits, one addressable by the contacts and the other by the antenna. This is the most secure, but also the most costly solution in the long term.

Other designs allow the same microcircuit to be addressed by either method. Either dual-ported memory is used or the same microprocessor supports both contact and contactless interfaces (see Figure 7.2). In this case it is essential that the logic knows which route has been used and controls access to memory accordingly. This *dual interface* design may well be the preferred architecture for the future, but the devices have only

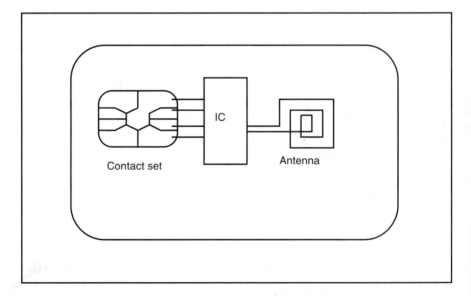

Figure 7.2   Dual interface *combi* card.

recently become available for widespread commercial use, and it is too early to say what their limitations will be.

Form factors

The very term *smart card* implies a card-shaped device, and we have come to take it to mean a credit-card size. However, smart-card technology is used in several applications where other shapes and sizes are preferred.

Modules

A smart-card module is more or less the smallest size of plastic card that can accommodate a smart-card chip and its contacts (see Figure 7.3).

Modules are used as *security application modules* (SAMs) within readers and terminals; a SAM contact set can be mounted directly onto a printed circuit board and in some applications the module is locked in place by *potting* the relevant parts of the reader (fully enclosing them in a resin compound that cannot be removed without destroying the circuitry).

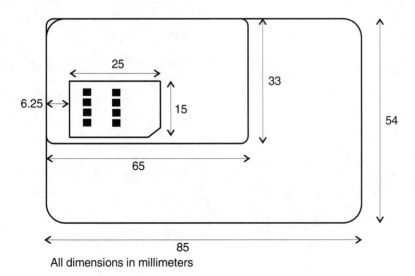

All dimensions in millimeters

Figure 7.3 Module and minicard sizes.

They are also used in mobile telephony, as *subscriber identity modules* (SIMs) in *global system for mobile communication* (GSM) telephones and in some portable instrumentation.

Minicards

Minicards (see also Figure 7.3) are a size between the module and the ISO 7810 credit card. The slightly smaller size is potentially cheaper to manufacture, and the card is more rigid and less liable to break due to mechanical stress.

Minicards are being proposed for several public-transportation applications, where they often correspond to the size of existing magnetic-stripe tickets. Because of this, there has been some difficulty in agreeing on an international standard for minicards, with each country arguing for the size of its existing tickets.

Figure 7.4   Smart diskette. (Photograph courtesy of Fischer International Systems Corp. (U.K.) Ltd.)

Diskettes

Where smart cards are to be read by or used in conjunction with PCs, there is an advantage if they can be read by existing devices rather than requiring an additional reader. Fischer International produces a device based on the smart-card chip, but which draws its power and communication from the magnetic field of a floppy disk drive (see Figure 7.4). This proprietary device can therefore be used for security and logical access control on a PC system.

Keys

Smart-card chips can also be built into key-shaped devices (see Figure 7.5). These are physically very robust and can also be made with a mechanical action in addition to their logical functions. The contacts appear either in a row or as a series of rings around the shank of the key.

Figure 7.5 Smart key. (Photograph courtesy of Nexus (G.B.) Ltd.)

Others

Many other shapes and sizes are also possible, provided only that the carrier is large enough to support the microcircuit. Many of these proprietary devices gain security from their proprietary nature: They will usually be uneconomic to reproduce or to hack into. For high-security applications, however, they should be treated with some caution, as none of these devices incorporates the specialized techniques or years of development that have gone into the high-end products in the standard shape and size. Speed, memory capacity, and special features are all sacrificed in the majority of proprietary designs, and the smaller manufacturers are also unlikely to have the extensive controls and procedures in place to protect the manufacturing and personalization process fully.

8

Smart-Card Components

Carrier

As with magnetic-stripe cards, the basic material of the smart card is polyvinyl chloride (PVC) or a similar thermoplastic. PVC is most commonly used for international bank cards because it is cheap and can be embossed. Acrylonitrile butadiene styrene (ABS) is also widely used; this slightly more brittle plastic cannot readily be embossed, but it will withstand higher temperatures and thermal cycling. This allows the chip to be bonded to the card at a higher temperature (with PVC cards the module is usually glued in to the card). Polycarbonate also has a higher temperature specification and is used in mobile-telephone cards, which call for operation at up to 100° C. Polyethylene terephthalate (PETP), which offers a very flexible and light card, is used widely in Japan.

The card production process will be discussed in more detail in Chapter 10; however, with all of these materials the chip and its contacts are normally embedded in the card in the form of a module, as described in Chapter 10. The module is bonded into a milled or molded indentation in the card body, usually after the card has been printed.

Several manufacturers have variations on this process, each with its claimed advantages. At least one manufacturer uses polycarbonate as the carrier material in a shortened card production process; the chip is inserted directly into the card and the contacts are printed onto the surface of the card using a conductive ink as a part of the printing stage. Another manufacturer has shown that it is possible to make cards using wood as the carrier material.

No card material is inherently superior to all others, although some may be better for individual applications. Card security is improved slightly when the chip or module cannot be removed from the card after manufacture (this would make some forms of circuit analysis easier); with resin-bonded modules this is sometimes possible.

What is important, however, is the life of the card. It is the mechanical features of a smart card that determine its life; failure or unreliability usually arises from contacts becoming damaged or the links from the chip to the contacts being broken. This can come about as a result of mechanical stress (particularly twisting) or thermal stress (differences in expansion coefficients between the plastic and the wires). In applications where high reliability is required, a plastic that deforms little with heat and is not prone to fatigue failure with repeated twisting should be used. Of the widely used materials, ABS probably meets this requirement best, although polycarbonate may be as good or better.

External security features

Smart cards can incorporate several of the security features described in Chapter 4, including:

- Microprinting;
- UV sensitive ink;
- Hologram;
- Signature strip or printed signature;
- Cardholder photograph.

These features will often be less relevant in the case of a smart card (which is designed to be authenticated electronically) than for a magnetic-stripe card, for which visual inspection is usually the only form of card authentication.

Bank-issued smart cards will also often incorporate a magnetic stripe so that they can be read in countries or by retailers that are not equipped with smart-card readers. It is a rule of the international card schemes that a retailer who accepts, say, Visa cards must accept all valid Visa cards, not just domestic cards.

Cards may also carry a small barcode in one corner; this is a batch control used in manufacturing and personalization, and does not form part of the operational characteristics of the card.

Chip

The key component of the smart card is the integrated circuit embedded in it. As we described in Chapter 7, this may be a memory, protected memory, microprocessor, or FPGA chip. Several semiconductor manufacturers now make ICs specifically designed for smart cards, generally using:

- *Low-power CMOS technology,* operating at voltages between 2.2V and 5.5V: chips operating down to 1.5V and below are being designed;
- *Very small feature size* (the width of the smallest line or separation on the chip): 1μ is now common, and the most advanced IC production lines are now working at 0.2μ. Smaller sizes still are in view;
- *Special manufacturing controls* described further in Chapter 10.

These chips are normally available from the IC manufacturer in the form of sawn wafers, tapes, or modules for direct embedding in cards.

Microprocessor

Most current smart-card micros are 8-bit, usually based on Motorola 6805 or Intel 8051 designs, and with clock speeds up to 5 MHz. Sixteen-bit processors are used in some newer chips, and will probably become more common in security applications. Higher clock-speed chips are also now being produced.

A small number of *reduced instruction set* (RISC) microcontrollers are now available for use in smart cards. These will be particularly useful in applications that require high speed over a limited range of functions.

Memory

Memory is the largest variable item in the design of a smart card IC. The memory is first divided into areas according to the type of semiconductor memory:

- *Read-only memory* (ROM) is used for storing the fixed program of the card, also known as its *mask*. The mask is sometimes thought of as the card's operating system, but often it is the only program. Part of the ROM memory (known as user ROM) may also be available for what we can think of as application programs running in the card. ROM is efficient in both space and power requirements.

- *Programmable read-only memory* (PROM) is used for loading a card's serial numbers or other fixed values. This part of memory is usually very small (up to 32 bytes) and accounts for little space or power.

- *Flash* memory is sometimes used for storing additional programs for execution within the card, as well as for storing blocks of data which will be read and extracted all at the same time. This is because flash memory can be written using the normal card communications interface, but it can only be erased as a single block. Flash memory is more efficient in space and power than E^2PROM, but less so than ROM. It is not widely used in smart cards today, although it is becoming more common.

- *E^2PROM* is used for most data storage—the smart card's equivalent of the hard disk on a PC. It can be read or written at any time (subject to the security logic), but is very costly in space and in power. For many smart-card chips, the E^2PROM accounts for more than half of the total chip area, as shown in Figure 8.1. Two important parameters for the life and reliability

Figure 8.1 Smart card microcomputer chip. (Photograph courtesy of Siemens AG.)

of chip cards are the *number of write cycles* the E^2PROM can reliably accept and the *data retention period* (the period for which the memory will reliably retain data). Although the quoted figures for these are usually quite high (typically 10,000 cycles and 10 years, respectively), both are statistically determined (they depend on the rate of migration through the semiconductor junction) and in a large card population some cards will start to show unreliability well before the quoted limits.

- *Random access memory* (RAM) is used for temporary working storage. Data in RAM is lost when the card is removed. Whereas PCs and other small computers normally have access to a megabyte or more of memory, smart-card micros typically have memory sizes from 128 to 512 bytes.

- *Ferro-electric RAM* (FERAM or FRAM) is starting to be used in smart cards. This type of memory consists of RAM with an additional layer, which has the effect of making it nonvolatile (i.e., it retains its memory without power). It can therefore be used in place of E^2PROM, over which it has two major advantages:

- Write time is the same as read time: nanoseconds compared with milliseconds for E^2PROM;
- The power required for writing is also much less than for E^2PROM; this is particularly important for contactless cards because it reduces the total power requirement of the chip.

FERAM has a much higher packing density than E^2PROM and an perform as volatile or nonvolatile memory. This makes it possible for a manufacturer to design the whole of memory as a single block, designating areas as ROM or RAM at customization time. One drawback today is that it is a proprietary technology; although it has been licensed by several companies, the final layering is always carried out in the United States.

Different applications require varying proportions of these different memory types, and it follows that cards are divided into applications according to these ratios, as well as by their power and the functions built into the mask.

For example, a card with a large flash memory may be used for storing a history of operations, for storing health records, or for telemetry. Applications such as bank cards require a large amount of ROM because of the complexity of the programs, whereas the new generation of security modules for mobile telephones requires very large E^2PROM areas to store subscriber profiles as well as passwords, phone books, and call history.

Coprocessors

Smart-card chips that are designed for use in security applications very often have a built-in coprocessor. In some cases this processor carries out standard multiple length arithmetic operations (multiplication and exponentiation) in hardware; in other cases the coprocessor is designed to execute common cryptographic functions, such as DES encryption or RSA signatures, directly.

Memory management

The memory of a smart-card chip is controlled by memory-management circuits, which provide hardware protection against unauthorized access. This protection makes use of a hierarchy of data files: the highest level is termed a *master file* (MF), below this may be several layers of *dedicated file* (DF), and finally one layer of *elementary file* (EF) (see Figure 8.2). Access to a lower level must pass through each of its parent files in turn; this forms a logical channel to the elementary file. Parent files contain the access rules and control information relating to the next layer down. ISO 7816-4 (see the appendix) defines the security mechanisms appropriate to provide both authentication and confidentiality of the data, keys, and control fields during normal access. The card itself provides these security mechanisms, and the designer must ensure that all fields are correctly identified and that the structure of the file hierarchy is appropriate for the application.

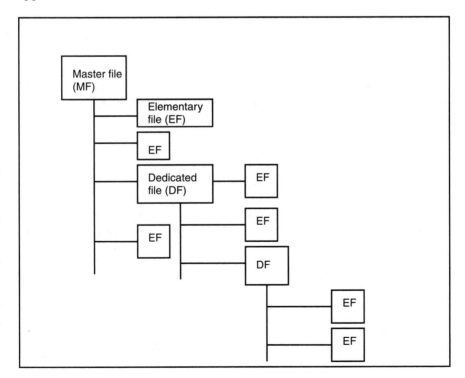

Figure 8.2 Smart-card file structure.

Input-output

All microprocessor smart cards make use of a single bidirectional serial input-output interface. As we will see, the electrical connection may be through contacts or by electromagnetic coupling. The standard for this communication is set by ISO 7816-3, which provides for different optional communications protocols: character- or block-oriented, synchronous or asynchronous. Nevertheless, the vast majority of today's smart-card microprocessors use the asynchronous character-oriented protocol defined in ISO 7816-3 as $T=0$.

The speed of communication is set by a combination of the clock speed (defined by the terminal device) and the dividers in the card. The terminal "wakes up" the card by applying power followed by a reset (RST) signal. The card responds with an answer to reset (ATR), which tells the terminal what type of card it is and which communications protocol it will use.

The input-output interface for a contact card consists of two connections: I/O and GND, supported by an optional clock signal.

Chip security features

In addition to these features, which are designed to protect the data inside the card during normal operation, the security of the card also requires that no one can read the data other than by sending commands to the card. A number of attacks can be envisaged.

Use of test modes

During card manufacture, test modes and test contacts are provided on the chip. Where these are provided, they are invariably disabled (by blowing a fusible link) once the tests have been successfully completed.

Electron microscopy

Electron microscopes inspect the physical structure of the chip. To protect against this, the buses may be scrambled or the logical memory structure may be mapped onto physical memory in an apparently random way: One bit is stored here and another there, so that without access to the memory mapping the analyst cannot reconstruct the data. Some randomization is used on all but the simplest smart-card microprocessor chips. More complex chips are constructed in layers, with the memory-management functions, PROM, and other highly sensitive elements in the lowest layers.

To protect further against electron microscopy, security chips may include a *metalization layer* over the whole chip; this in turn includes some links into the main circuitry so that any attempt to remove the metalization layer destroys the chip. Some high-security chips also include circuits that can detect the radiation of an electron microscope and destroy all the data in memory (E^2PROM data is likely to be erased by the beam in any case).

Circuit analysis

Circuit analysis involves low-frequency manipulation of the inputs and observation of the outputs. Many chips include detection circuits, which detect any attempt to operate the chip outside its normal clock or timing parameters and can shut the chip down.

Measurement of operating parameters

Many smart-card chips are quite slow compared with other computers. It is therefore possible to measure the time it takes to perform certain tasks and make deductions concerning the results. For example, a "hit" in a search table will often produce a shorter operation time. Or the current can be measured, yielding information about the quantity of data being written. Millions of repeated observations of this type can yield useful data for analysis. Some chips therefore introduce null operations, or random variations within a single operation, to destroy the usefulness of any information that may be gained by external measurements.

Deliberately induced faults

It has been proposed that data, particularly the secret keys of a public key cryptogram (see Chapter 5) could be extracted by inducing an error in an iteration (preferably the penultimate) and observing the result of the last iteration. This could be done by operating the chip outside its temperature range or by radiation. Although the chip should ideally cease to operate as soon as its operating range is exceeded (as it would in the case of a clock or voltage variation), very few chips carry detectors for temperature and still fewer for radiation.

In practice, however, this type of error could not be introduced systematically, and the observer would therefore have to analyze an exceptionally large quantity of data with no certainty of obtaining a result—a level of effort that would be disproportionate to the rewards in virtually all cases.

The standard protection against this attack is for the smart card to authenticate the terminal before commencing any operation. The chip can additionally protect against such attacks by performing operations twice or by reversing a cryptographic transform and checking that the original value is obtained. Although these lead to increased operation times, they will provide the necessary security in those applications in which it is paramount.

Contacts

A normal (contact) card has up to eight contacts, the position and designation of which are defined by ISO 7816-2 (see Figure 8.3). Many French cards still use a different contact position, which is designated in the ISO standard as "transitional." Although there is a clear intention to phase out this older contact position (in the top left-hand corner of the card), a large number of terminals in the field will accept only this older contact position, so there will be a long delay before conversion is complete.

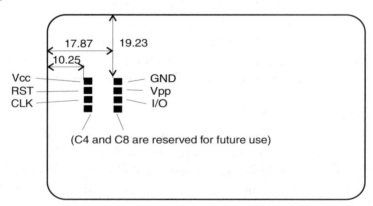

All dimensions in millimeters

Figure 8.3   ISO 7816-2 contact positions.

The contacts may be made of gold or another conductive material; they are usually connected to the chip itself by very thin wires during the module manufacturing process. Because of the flexibility of the card, these connections are a potential source of unreliability. The contacts themselves have a limited life, particularly when they are used with cheap readers using *sliding* contacts.

Antenna

In a contactless card, all input-output—and often the power—is transmitted by *radio-frequency* (RF) signals rather than through contacts. A coil antenna is built into the thickness of the card, either around its circumference or with a larger number of turns around the chip itself, within the size of the module.

Some contactless cards use a battery within the card. In others, the RF-control circuits in the chip pick up sufficient energy through the antenna to power the chip itself and to communicate with the read-write unit. The signal uses frequencies designated for telemetry: 125 kHz or 13.56 MHz. The card communicates by modulating the RF signal, using one of the standard modulation types: *amplitude modulation* (AM), *frequency modulation* (FM), *amplitude shift keying* (ASK), *frequency shift keying* (FSK), or *binary phase shift keying* (BPSK). Card designers try to use the system that will require the least power; a protocol such as *code division multiple access* (CDMA), which would yield the most secure communication, requires too much power for today's card antenna/chip combinations.

Power requirements drive contactless designs; typical microprocessor cards require 5–8 mW of power, while 1 mW–1.5 mW may be adequate for small protected memory cards. The power required increases as a cube of the distance from card to reader. More power can be transmitted at the lower frequency, but data rates are lower. With the lower frequency, the distance from the card to the read-write antenna can be up to 1m, with the higher frequency the maximum is usually closer to 20 cm. The maximum data-transfer rate, however, increases with frequency (up to 100 Kbps for the 13.56-MHz cards).

In many applications, the actual reading distance is kept smaller than the maximum possible. The larger the reading distance, the more likely it is that there will be several cards in range at one time; thus, the chance of communications collisions is greater. In financial applications (even where a bus fare is being deducted from the card) users like the more positive action of bringing the card into contact with the reader, rather than having value removed from the card while it is in their pocket. For other applications (building-access control or season tickets in public transportation) the ability to read at a distance is a positive advantage, and for these applications some form of anticollision protocol is required. These protocols have been designed for a wide variety of networks and do not pose any problems to the system designer.

Contactless smart cards are able to make use of any of the security mechanisms (controlled memory access, card and reader authentication, data encryption, etc.) available to contact cards. Some of the lower end memory products do not make use of these mechanisms, and they therefore raise the specter of data being read from a card, or a transaction effected, without the knowledge of the cardholder.

Where card and reader authenticate each other, the only question is whether a second reader could gain any useful information from passively listening in on an exchange between a valid reader and the card. This is probably more of a theoretical possibility than a practical likelihood, but if the eavesdropping device had knowledge of the public keys of the two devices, there is a possibility that it could extract a session key (or any other symmetrical key) from the exchange and thereby decode data encrypted using that key. For systems requiring high security, therefore, key distribution within a contactless card scheme should only be carried out in a secure physical environment.

Mask

The fixed program of a microprocessor card (often referred to as its operating system) is held in ROM, and it is called the card's mask. Different versions of a card can be produced on the same microprocessor chip by using different masks. Unlike a general-purpose computer, for which the operating system only carries out such functions as file management, scheduling, and intertask protection, the mask of a smart card often performs application functions: reducing the value in a purse, for example, or comparing a digitized signature with a stored pattern. Some of the application logic may be contained in the mask: A smart credit card, for example, will not complete a transaction until the terminal has executed the correct sequence of checks.

Smart-card microprocessor manufacturers can supply a default mask, including the minimum functions required to make use of the card's features. This is often built into the development kit for the card. Most smart-card manufacturers (which take the module and build it into a card) have developed their own masks for popular applications, as have several card personalization companies (which are principally concerned with those elements that can be loaded after the card is manufactured).

As the mask must be programmed into the chip when it is manufactured, its security forms part of the manufacturing process. For high-security chips, the manufacturer insists on a tight control of the development tools and process as well. These processes will be described in Chapter 10.

From a security point of view, a microprocessor card operating system consists of a set of layers, as shown in Figure 8.4. The innermost layer is concerned with handling access to individual fields and how those fields may be manipulated. The file manager deals with the DF-EF structure and the access rules for those files. The command manager

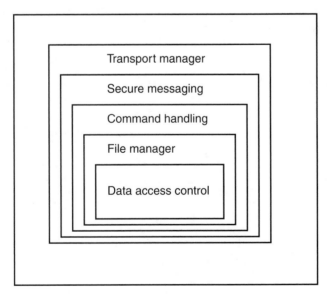

Figure 8.4   Card operating system structure.

Table 8.1 Sample Card Specifications

Application	Manufacturer	Processor	ROM	PROM	RAM	E²PROM	Coprocessor	Security Features Included
Phone card	ODS (Siemens chip)	—	16 bits	56 bits	—	32 bits	—	Security logic; count down only
Eurochip (2nd-generation public telephony)	Philips	6805	6 KB	—	128 bytes	1 KB	—	Randomized response times; low frequency protected
GSM telephony	SGS-Thomson	6805	16 + 1.5 KB	—	384 bytes	8 KB	Arithmetic	Address scrambling
Bank (credit / debit)	Gemplus	6805	8 KB	—	—	1–8 KB	Arithmetic	—
Health/secure payments	Solaic	8051	8 + 2 KB	—	256 bytes	2.5 KB	Crypto (RSA)	—
Computer access	Philips	8051	20 KB	—	640 bytes	8 KB	Crypto (fast exponentiation)	Low frequency and voltage sensors

interprets commands and checks whether that command is valid at this point in the operation. The next layer implements the secure messaging defined in ISO 7816-4, which is designed to provide both authentication and confidentiality of the data exchanged between reader and card. Last, the transport manager provides the low-level communications functions required by ISO 7816-3. This structure provides a very high level of confidence in the ability of the card to protect the data stored in it, under all normal operating conditions.

Reliability factors

All of the elements described here have an effect on the reliability of the card. The carrier must, as we have seen, have very good strength and flexibility, without imposing undue stress on the chip itself or on the connection to the contacts. The contacts themselves must be thick enough to last the life of the card, bearing in mind the range of readers in which it may be used.

The chip size is very largely dependent on the area required for E^2PROM, as this takes up the most space. It is usually argued that a smart-card chip should not exceed 20–25 mm^2; above this size the chip itself becomes too fragile. The amount of E^2PROM which can be accommodated in this area depends in turn on the feature size, which is linked to the operating voltage. Current smart-card technology, with feature sizes of 0.7μ–1μ and voltages of 2.7V or 3.3V, can accommodate up to 8 kb or so of E^2PROM within this area.

Although greater complexity will in principle reduce the reliability of the card, in practice the security and other design features used in the more complex cards result in a unit that is at least as reliable as its simpler brethren.

Sample card specifications

Table 8.1 gives some examples of card specifications for different applications. The security features mentioned are examples only; all cards will have several features not mentioned here. The table does not include cryptographic functions, which are described in Chapter 5.

9

System Components

A SMART CARD is an element in a distributed computing system. Having no user interface of its own, it can only operate as a part of a system. The security of any system is only as strong as its weakest link; thus, even if the smart card itself were totally impregnable, the data could be compromised by other parts of the system or by the processes and procedures that make use of the card. This chapter is concerned with the other system elements, and in Chapter 10 we will look at the procedures.

Reader

Closest to the card is the card reader. This is sometimes called a *read-write unit* (RWU), *write-read unit* (WRU) or card-accepting device because it can write to the card as well as read data from it.

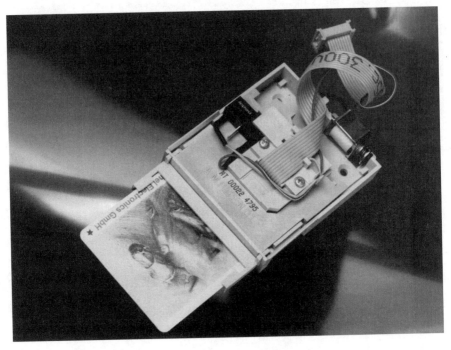

Figure 9.1   Chip card acceptor device. (Photograph courtesy of Amphenol-Tuchel Electronics GmbH.)

Contacts

The card reader for a standard contact card consists of a set of contacts, some logic or intelligence that acts as an interface between the card and the rest of the terminal, and the card transport mechanism.

The basic element of this reader is the set of between five and eight contacts which establish the electrical connection with the card (see Figure 9.1). These contacts may be sliding (the card is moved into place under the contacts) or landing (the contacts only close onto the card once the card is in place). Sliding contacts have a slightly higher chance of making a good connection, but at the price of a greater insertion pressure and greater wear on the card contacts. It is also more difficult to ensure that power is removed from the card immediately if the card is withdrawn; if this is done too slowly there is a possibility of giving wrong results or of feeding the wrong voltages to the other contacts. The best contacts have a small amount of sliding as they land.

Card transport

Readers for smart cards may be of the manual insertion type or motorized. Both have implications for system reliability and security:

- Insertion readers are much cheaper: They have few moving parts and require very little power. They are, however, more easily vandalized (by inserting a metal plate, screwdriver, chewing gum, or even fast-setting cyanoacrylate adhesive). Card capture is less important for a pure smart card (because the chip itself can be disabled by the reader), but most bank cards will continue to have magnetic stripes as well for some time. Insertion readers are generally unable to capture cards. And there is a higher risk of the card being withdrawn before the transaction is completed; as we will see, this requires specific logic or software to bring the card and system back to a known and consistent state.

- Motorized readers (see Figure 9.2) are able to avoid the vandalism problem to a large extent by using a shutter (although on many existing readers this is opened by detecting the magnetic stripe). They are able to capture cards, and the card is fully under the control of the system while the transaction is in progress; this leads to a more consistent operation. On the other hand, motorized units require more power and, having more mechanical parts, must be made more robust. Hence, it is considerably more expensive to achieve a given reliability.

The card transport must also be able to provide a reliable indication of the reader status: in particular, where there is a card in the reader, and if it is fully inserted. The standards require power to be removed very quickly from the card if it is withdrawn prematurely.

Control electronics

Behind the read/write contacts there must be some intelligence, and some of this is usually provided within the reader itself. As a minimum, the reader circuits must handle the power-up sequencing, card transport, and status signals. Memory card readers will usually provide a simple command-driven interface to the host system, while readers designed for

Figure 9.2 Motorized reader. (Photograph courtesy of Datastripe Ltd.)

microprocessor cards will also control the card timing and buffer the communications.

For security applications, the reader circuits must often go further than this, by providing some integrated authentication or encryption functions. For the system to be secure, the exchange of any authentication data must take place in such a way that it cannot be intercepted; this normally means that it should take place as close to the card as possible. The reader itself, and the authentication electronics, must be tamper-proof, or at least tamperevident. This is usually done by *potting* the electronics in an epoxy resin and ensuring that if the housing is opened, the memory is erased.

If this is not done, then all communications between the card and the authenticating system must be encrypted; not only this is a major over-head, it also raises difficult questions as to how to prove the integrity of the encryption key used in this communication.

Where the identity of the cardholder is to be verified by the card itself (by checking a PIN, a signature pattern, or other data), it must also be impossible to intercept the data being passed to the card (because this would give away the secret data stored in the card). It is therefore normal to insist that the keypad, signature tablet, or fingerprint unit are a part of the same tamperproof housing, or else that the data be encrypted. As it is often not practical for the whole unit to be made tamperproof, the reader electronics must be able to perform DES encryption according to ISO 9564 standards.

Where keys must be stored securely within the reader, or where data must be signed using a private key for authentication (see Chapter 5), a SAM is often the best way to protect the key. A SAM is a smart-card module, usually incorporating a cryptoprocessor and memory, which forms part of the reader electronics (see Figure 9.3).

Figure 9.3 Reader electronics and security application module. (Photograph courtesy of Landis & Gyr S.A.)

Contactless-card readers

Card readers for use with contactless cards are able to dispense with the contacts and card transport; in place of these they make use of an antenna, which is usually wound as a coil and mounted behind a plate. The reader can be mounted in whichever position is likely to be most convenient for users and is likely to give the best coupling with the card; for public-transportation applications it will often be located at waist height, to allow the card to be carried in a pocket.

In addition, the electronics must control and decode the RF signal. This will often require logic to detect multiple cards in the field and to ensure that communication only takes place with one card at a time. As we described in Chapter 8, this logic is not as secure as that used by the best military and commercial radio systems; this is one reason why contactless cards are still not used in applications demanding the highest levels of security.

Terminal

The reader will be built into some form of terminal—a free-standing device that connects to a network. These terminals may take several forms, such as PINpads, PC readers, *electronic point-of-sale* (EPOS) or *electronic funds transfer at the point of sale* (EFT-POS) terminals, ATMs, vending machines, and access control.

PINpads

A PINpad is a small handheld terminal comprising a keypad, display, and smart-card reader (see Figure 9.4). Its purpose is to allow the cardholder to enter a PIN without having to disclose that PIN to the shop assistant or to anyone else. PINpads therefore often have some form of cover over the display, and sometimes even the keypad, to stop others from observing the PIN as it is entered. The design should take into account the physical layout of the environment in which it will be used: Would it, for example, be possible for someone to be looking over the shoulder of the person entering the PIN ("shoulder-surfing")? Since most keypads conform to one of two standard layouts, someone using binoculars or a telephoto lens can observe the operation from afar.

Figure 9.4 PINpad. (Photograph courtesy of ICL Financial Terminals AB.)

The PIN must also be transmitted securely to the card; if the card reader forms part of the PINpad, then the whole terminal can be made tamperproof. System designers should ensure that if the terminal is opened (for example to insert some additional wiring to monitor the operation), it is not subsequently possible to reactivate the terminal without its passing security controls.

Where the card reader does not form an integral part of the PINpad, the PIN must be encrypted as it is passed from the pad to the card.

PC reader

Smart cards are increasingly used to give access to computer systems, and data must often be loaded onto cards using a PC. In the world of electronic commerce, as we shall see in Chapter 12, smart cards are likely to play an important part in providing secure payments across public networks. For all of these purposes, smart-card readers must be attached to PCs.

A typical PC smart-card reader fits into a floppy disk drive bay, and may communicate with the PC through a standard serial port or a special bus card. The serial port is appropriate for reading and writing memory cards, and for situations in which there is a high level of intelligence built in to the card reader itself. Where more of the functions are carried out on a standard PC board (with simpler reader electronics), the security of the interface between the reader and the PC board must be very carefully checked, as this is a potential weak link.

EPOS or EFT-POS terminal

Most large retailers in northern Europe and North America now make use of electronic point-of-sale (EPOS) terminals. In addition to the functions of the standard electronic cash register, the EPOS terminal is connected to a network and can be programmed for additional functions. The cash register has access to a database of items in the shop; as goods are sold, their barcodes are scanned and the purchase is recorded. Full details of the transaction are sent to a back-office computer, and selected data are sent to the head office for further analysis.

Where smart cards are to be used for recording the payment or customer details alongside the transaction, a smart-card reader or PINpad must be attached to the register or terminal. EPOS manufacturers have not yet developed fully integrated functions for smart-card reading; thus, at this stage it is likely that any reader will have to have a sufficiently high level of intelligence to carry out all the smart-card-specific functions; the EPOS terminal acts as a network component. In due course, software modules will be developed to carry out these functions in the terminal, but this then raises the problem of ensuring the integrity of the smart-card application if changes are made to other parts of the software.

The solution to this problem lies in the inherent security of the operating systems and the link between the EPOS terminal operating system and that of the card. Several companies are currently working on operating system and terminal structures to provide this mutual security, but no commercial products are yet available. As we will see, this problem has been addressed more successfully by ETSI and the telephone companies.

Another solution, which offers better security in the short term but has operational disadvantages, is to separate the payment function into a stand-alone electronic payment—or EFT-POS—terminal. This provides the functions of a PINpad, but with the addition of a printer, modem, and software.

ATM

The next form of reader is the automated teller machine (ATM). All ATMs today are motorized and driven by magnetic-stripe cards (even in smart-card countries such as France). The issue here is to provide smart-card reading as well as magnetic-stripe reading, as a single upgrade. ATMs are always connected online to the bank network to perform transactions and include symmetric data encryption in their software. But the smart-card reader must be able to perform the card authentication (an asymmetric algorithm) and other intimate functions before passing the transaction to the host system.

Vending machine

Many vending machines have already been adapted to accept smart cards; the operators find that the benefits of reduced vandalism and not having to collect cash far outweigh the extra cost of the card. Most of these machines use proprietary cards and interfaces, and their security depends not only on the conventional measures discussed elsewhere in this book, but also on the control exercised by the manufacturer over the technology.

As smart cards become more pervasive, vending machine operators and their customers will want to make use of stored value cards issued by others, including electronic purses offered by banks. The vending operators will have to adapt their systems to accept cards meeting the open standards; many have already started to move in this direction.

Most vending machines accept only one form of payment. There is usually only one operation, which also lends some security to the system. Typically, insertion readers are used, and the transaction data is stored in flash memory until it is collected by a host system. It is not usually necessary to store details of the card used, although provided that the data are stored in an encrypted form there are few security implications in storing the card number as part of the transaction.

The use of smart cards will permit increasingly complex and higher value transactions to be performed through kiosks and vending machines, including ticket issue for public transportation and events, sales of holidays and insurance, and ordering for home delivery. In these cases, it is likely that a bank-issued card will be used, and the kiosk must incorporate the full facilities required for card authentication, cardholder verification, transaction encryption, and secure storage until the unit is polled by the host system. Many will also need to be able to connect online to a bank host to authorize the transaction amount.

Access control

Smart cards are not yet widely used for access control. There are several reasons for this:

- The action of reading a smart card is slower and more awkward than swiping a magnetic-stripe card;
- The extra cost is rarely justified unless other cardholder-verification techniques (biometric or PIN) are used at the same time;
- Access-control companies have concentrated their development on online systems.

This could change as contactless smart cards become more wide-spread. The contactless card is likely to replace the traditional Wiegand tag in many security-conscious establishments. Smart-card readers for access control are therefore most likely to consist of aerial pads located adjacent to doors, although in some military or high-value financial situations the extra safety of a contact card combined with a biometric check will still be preferred.

Others

Free-standing smart-card terminals, often similar to a PINpad but with additional intelligence and memory, are used in a variety of applications, including retail-loyalty schemes, gaming control, and offline functions such as loading and balance checking in closed loop systems.

Stored-value schemes often require simple balance readers, which can be made available to the cardholder. These are usually in the form of a key-fob, just large enough to take the corner of the card and to display a value on a single-line LCD display.

Multicard operations (e.g., transferring money from an electronic purse onto a membership card to pay a subscription) will require some form of electronic wallet, and in some models of electronic purse this will also allow value to be exchanged between individuals. The wallet looks like an electronic calculator, with the addition of a smart-card reader slot.

Terminal protection

Most terminals must be physically protected. If they are not in a secure area, then we may need to consider regular polling or dialing in, and the terminal should incorporate self-diagnostics so that it can inform the central system of any malfunction or attempt to tamper with it.

Network

The role of the card

The card, as we have said, is only one element in a distributed computer system. In some cases, it is providing data to the system; in other cases, it allows access to a program or to data on the host system. Most card operations are transactions—the card and system exchange some data, agree that they are prepared to do business with one another, exchange details of the transaction, and then break the connection. Usually there is only one exchange of data between the card and the host—all the other functions are performed by the reader or terminal.

There is one special case in which the card "transaction" has a much longer duration, and this is when the card is used as a key to gain access to a computer system. In this case, the transaction lasts for as long as the online session and must remain in the reader for all of that time.

Network security checks

Where a card transaction must pass, for any part of its journey, through a network that cannot be regarded as fully secure, precautions must be taken to ensure that the purpose of the card—be it authentication, confidentiality, integrity, or any combination of these—is not compromised.

We must start by assessing which parts of the network are secure. We have to look at:

- *The system design:* Do we control all of the network? What protocols are used and do they provide the security we need? Are there backups if critical items fail? Are the backups equally secure? What other access points to the network exist?

- *Physical security:* Is the building secure? Is all critical equipment in a room with controlled access?

- *Organization:* Who has access to the system (at any level)? Who controls that access? Who has responsibility for critical elements? Does the staff policy meet the needs of security? If part of the network is subcontracted, does the contract include the security requirements and appropriate service level clauses?

These aspects are covered in more detail in the references in the bibliography.

Provision of network security

For those parts of the network that we cannot control, we must implement the security measures ourselves. For *reliability and integrity,* this will mean providing backup paths, message numbering and message authentication checks. For *confidentiality,* messages must be encrypted; the keys themselves must be exchanged by an alternative route or under cover of a master key that is rarely changed. For *authentication,* a hashed value of all messages must be signed and the signature checked by the recipient; this process should ideally be carried out in both directions.

Further details of these techniques are provided in the references listed in the bibliography.

The Internet

Transactions that take place over the Internet are a special case: The transmission control protocol/Internet protocol (TCP/IP) used offers a high level of resilience and alternative routing, but there is no protection against eavesdropping, alteration of the data, or nondelivery. These functions must be provided by the host and client systems, which are usually independent of each other.

Many of the common clients, particularly the *browsers* and email clients, have implemented secure messaging schemes in conjunction with host software companies and credit-card schemes. The most common browser has an optional *secure sockets layer,* which replaces the standard operating system interface. This uses a version of the RSA algorithm, but with a shortened key to meet U.S. government requirements. The current schemes are, however, optimized for the transmission of clear text input at the keyboard, particularly credit-card numbers, and do not take account of the functions available through smart cards. This is, however, a fast-moving area, and several more comprehensive schemes are under development.

Fallback and recovery

The network design must also take into account the need for fallback and recovery. Failures will occur, and some of these will happen during a transaction. When this happens, the two ends of the system often end up out of synchronization, with the data updated on the host system but not yet on the card or vice versa. The design of a smart-card scheme must provide a method for canceling a transaction that has not been completed. This is fairly simple for a single-stage transaction, but can be quite complex for operations that require changes to several different datasets.

Hacking

The community of hackers includes people with many different motivations, ranging from financial gain or industrial espionage to the enjoyment of technical skills and prowess. Any system with specific pretensions to security is a target for hackers who seek a challenge. Hackers do not usually have access to the massive computing resources required for brute force attacks on cryptosystems; they therefore seek easy ways in, particularly ways that involve learning passwords.

One of the specific advantages of a smart-card-based system is that it allows challenge-and-response authentication instead of simple passwords. Unlike a password, the response is never the same, and this defeats many potential attacks. Access control can also be by a combination of a token (the smart card) and a password or biometric: When challenged, the card asks for and checks the password before issuing the response. Neither stealing the token nor gaining access to the password on its own is sufficient to gain access to the system.

Host systems

It is not the place of this book to enter into the details of the security provided by host systems or of the measures which can be taken to improve that security. These subjects are covered in more detail in several of the references in the bibliography. In general, mainframe-based hosts already have most of the general security measures appropriate to systems of this type; smaller UNIX or Windows NT hosts are usually technically sound but often suffer from inadequate procedures and staff controls. In a smart-card environment, however, several additional areas need attention.

These mostly concern encryption and key management. Data that must be kept confidential should be stored either in encrypted files or files with formal access rights and regular security reviews. Keys must be stored in a secure manner: There are various techniques for this, but the most common method for secret keys is to divide them into components, with a different member of staff responsible for each component. They are then entered via a special device which recombines them to form the key.

These devices, which are known as host security modules (HSMs), come to form an important part of host system security (see Figure 9.5). As we shall see in Chapter 12, their use is obligatory in financial systems. They can perform encryption functions both online, under command of the host, or offline, via an attached terminal or PC. Key elements are held in semiconductor memory in tamperproof compartments. If they are tampered with or the power is removed, then the keys are destroyed.

Once a secure communication has been established between two parts of the system, keys can be securely stored on the host, encrypted under a key which is known only to the HSM. They can be transmitted

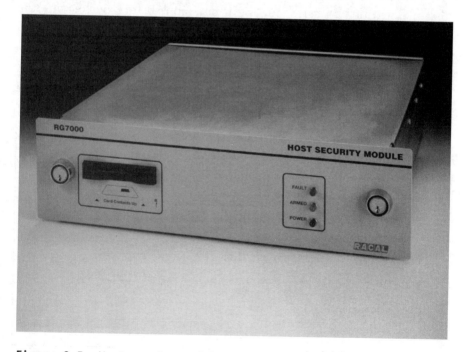

Figure 9.5 Host security module. (Photograph courtesy of Racal Airtech Ltd.)

to other parts of the system using the secure communications path, and operations such as digital signatures and signature checking are performed by the HSM itself.

Many smart-card systems are concerned with personal data; in this case the implications of data protection on the data stored must be kept in mind: In some countries, it may be necessary to encrypt all data to avoid unauthorized access, while in others the data must be stored in such a way that it can readily be accessed by cardholders and independent inspectors.

Trusted third parties

Authentication hosts

In many open networks, there is a need for one party to be able to authenticate the other (ensure not only that it is what it says it is, but also that it is entitled to carry out the transaction) before performing the operation. The medium for this is to use a *trusted third party* as an *authentication host*, with whom both parties are registered.

In Figure 9.6, user A is the card, which wants to check the identity of the host system B. A sends a challenge to B, which adds its public key, signs it, and sends it to the authentication host C for signature using C's

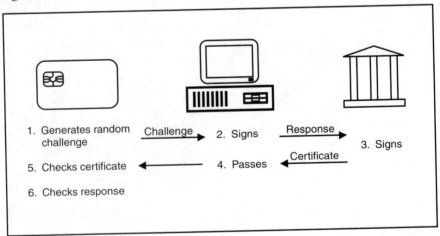

Figure 9.6 Authentication host operation.

private key. C returns the certificate to B, which forwards it to A. A checks the certificate (using C's public key, which is stored in the card), and if this proves genuine can then check B's signature. This completes the process.

This system, which is used by commercial schemes such as Kerberos, can become very cumbersome in a large network. A simpler scheme, used by the card-payment schemes among others, is for B to send its public key to C for certification in advance. When it receives the challenge from A, it is able to respond immediately, including the certificate in the message. This method requires some means of revoking certificates; the card schemes are able to do this because the certificates are stored in the terminal rather than at the card issuer host.

Evidence centers

Trusted third parties may also be used to store evidence, such as transaction certificates, and to perform checks on those certificates on request. In a large open system, such as a public electronic purse scheme, this may be a very important role, without which regulatory body approval would not be given.

Key escrow

The other role for which trusted third parties may be used is key escrow. Escrow is a legal term that covers the situation in which an item (or money or a design) is held on deposit by a third party as a contingency, usually against a breach of contract. In the case of key escrow, it is to enable the third party to decrypt information on request; this would typically be in the case of a legal investigation, fraud, or a regulatory check.

Key escrow has come into prominence because of national (primarily U.S.) government insistence on controlling strong encryption tools; as mentioned in Chapter 5, key escrow has been proposed as a route to allow such algorithms to be used in a controlled manner.

10

Processes and Procedures

M<small>ANY PROCESSES ARE</small> involved in the lifetime of a smart card; this chapter considers the security implications of each and how they are normally addressed.

Any system is only as strong as its weakest link, and nearly all computer system security breaches involve bad procedures or equipment breakdowns rather than failures of the security mechanisms themselves. Poor system design, or failure to enforce correct procedures, is the equivalent of installing an expensive lock and then leaving the key in the door.

Many systems designers are technologists and forget the importance of properly designed and documented *procedures*. The move towards ISO 9000 approvals has gone some way towards raising awareness of this very important subject, but too many organizations feel that because they cannot achieve ISO 9000, quality management is not important.

The fundamental requirement is to monitor the status of the card at every stage in its lifecycle. In this chapter we identify 12 or more

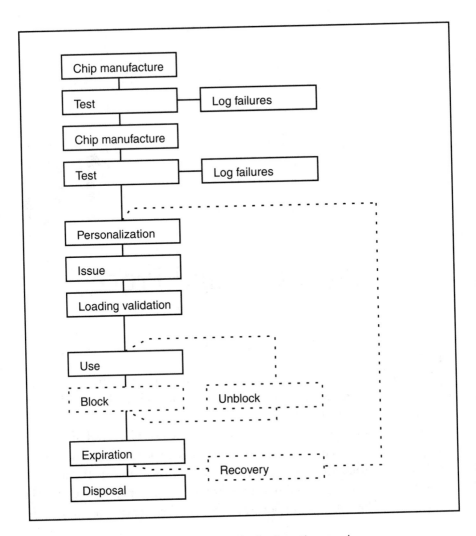

Figure 10.1   Card lifecycle for a single-function card.

states within the card's lifecycle, ranging from the early stages of manufacture and test to expiration and destruction; the number of states relevant in any given case may be higher or lower than this (see Figure 10.1). The card-management system must have provision for all the relevant states.

Chip design

The security of a chip card starts with the design of the chip itself. For commercial reasons, all design of large scale integrated circuits is carried out on secure *computer-aided design* (CAD) systems in facilities with tight access controls.

First the design is expressed in terms of design requirements (for different types of memory, I/O, and security). This is laid out in block diagram form and input to the CAD system. Further details of each block are then supplied to the CAD system, which as with all chip design starts to produce circuit layouts in the form of layers of materials.

A smart-card chip, whether memory or microprocessor, consists of a very large number of interconnected transistors. The layers represent p-type and n-type silicon, metal and oxides (insulators), with connections between the layers. After several iterations of this process, the design is complete and ready to be committed to photolithographic plates.

Already in laying out the circuits, the designers of security-conscious microcontrollers have a number of tricks available to make the analysis of their devices more difficult. Some of these were described in Chapter 8. They include:

- *Layering:* A complex microcontroller will consist of many layers of logic. In normal circuit design, it is regarded as good practice to keep like with like; to deter analysis, however, functions can be spread across several layers, in a seemingly random fashion. Elements regarded as vulnerable to analysis (particularly ROM) will be buried as far as possible in a lower layer.

- *Bus scrambling:* Buses may not be laid out in sequential order and may be in a different order on different layers.

- *Address scrambling:* Memory may not be laid out in the same order as its logical addressing would suggest.

- *Dummy components and functions* may be added where space allows.

- *Active components or links* may pass through or over the final *metalization* and *passivization* layers, which protect the chip against atmospheric effects and must usually be removed for any intrusive analysis.

The chip design may also include several special functions for the detection of attempted analysis or hacking, such as:

- *Low- and high-frequency detection:* This detects attempts to operate the chip very slowly for analysis or to introduce short pulses which might induce errors.

- *Temperature detectors:* These can shut down the chip when it approaches its maximum temperature. This is as much for reliability as it is an antihacking feature.

Few if any chips actually have all these features. All cost money, and most smart-card schemes are sensitive to the cost of the cards. The simpler cards typically used in telephone cards or closed user group schemes have very few of them. Even when the card has a full set of features, not all are used: Sensitive detection of clock speed or voltage can affect reliability, and features such as these must be managed.

Manufacture

The manufacturing process also starts with the chip (see Figure 10.2). The semiconductor manufacturer produces wafers of silicon carrying a large number of chips. These are divided into individual chips, tested, and then mounted either on *modules* (small plastic cards including the contacts and very thin wire connections from the chip to the contacts) or on strips, depending how the next stages of manufacture will be carried out.

Different smart-card manufacturers use slightly different processes, often as a function of the card material they are using. Figure 10.1 shows a typical process, in which a standard PVC card is manufactured, printed, and has its magnetic stripe applied. A small hole is then milled into the card surface, and the module is then inserted into this hole and glued in place using a strong epoxy resin. The advantage of this method is that the chip itself does not have to pass through any of the earlier stages, which can involve higher temperatures and heavy stressing of the card.

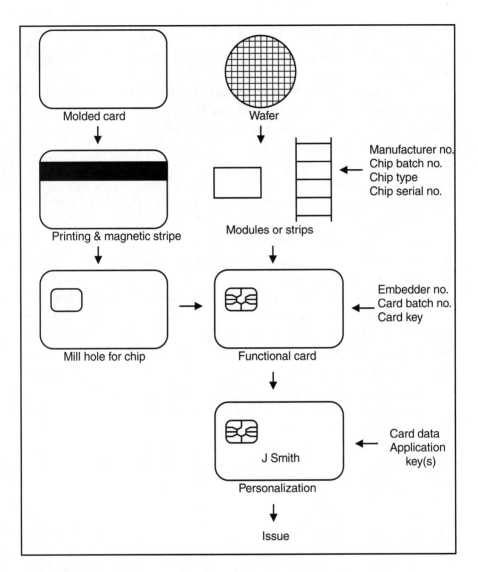

Figure 10.2 Typical smart-card production process.

The card itself is then tested and passed to the personalization stage.

Because smart cards are almost impossible to counterfeit, it is important to control theft of cards and components throughout the manufacturing stage. Again, strong emphasis is placed on good physical security in the premises in which chips and cards are manufactured. The resin glues used to bond the chip into the module are designed to be stronger than the silicon underneath, so that any attempt to remove the chip should destroy it.

The chip also passes through several clearly defined stages, and it is usually designed so that once it passes to the next stage it cannot go back. During manufacturing and testing, links used to provide access to testing functions or to load serial numbers and keys are blown after use.

During the initial mask loading, the semiconductor manufacturer (known as the *foundry* in the trade) writes data to the chip, which cannot be changed by subsequent operations. For maximum security, these data will include the manufacturer number, batch or wafer number, and component type. Once the chip has been tested, it is given a component number, and the *fabrication lock* is then blown before the chip is passed to the smart-card manufacturer (sometimes known as the *embedder*).

The card manufacturer embeds the chip in the card and often adds a further level of functionality (the *soft mask*) to the card, specific to the application. This process is also called *prepersonalization*. The card is tested again, and this time the embedder identification number, batch number, and card number are written to the chip. A card-specific key may be added at this stage. The prepersonalization process is completed by blowing the relevant lock, so that all of the data entered can never again be altered.

The card and chip have by now passed through several stages of testing. Products that fail any of these tests must be identified and controls must be in place to ensure that they have been destroyed.

Personalization

Personalization of cards may be carried out by the card manufacturer, by a specialist company, or by the user or scheme operator. Personalization involves adding the individual cardholder data; it is not required for an anonymous telephone card, for example.

Data transmission

We assume that the raw data is held on file by the end-user company. The first question to ask is whether the data to be stored on the card is itself sensitive: In this case it will probably be protected by some form of encryption or access control on the original system.

When the data is transmitted to the personalization company or department, the same level of security must be maintained. Some form of MAC or integrity check is probably desirable in all cases because it may not be possible to correct any wrong data written to the card. Where the data passes outside the company, a simple encrypting modem or a secure tape transmission service should normally be used.

Fixed and derived data

Some of the data may be added or calculated by the personalization company. This may include fixed data pertaining to the card issuer (the *issuer profile*). For multifunction cards, the data to be loaded in one phase of personalization may depend on data read from the card as a result of a previous phase.

Derived keys or certificates (see Chapter 5) should be generated inside the personalization device or a host security module attached to it. Similarly, where asymmetric keys are used and the private key is to be stored on the card, the keys should be generated by the personalization device. In this way, only the public key need be transmitted back to the scheme operator or host computer.

Testing

Once the data are loaded, the card is again tested. The tests may include functional tests (it is not always possible to test all of the card's functions before personalization), visual tests, and—very importantly—*synchronicity* tests, which ensure that the data written to the chip correspond to any data written to the magnetic stripe or printed on the card. Card personalization machines often have a carousel with several stations, which can allow cards to get out of sequence unless this check is carried out. Where cards will be issued by mail, the mailers are usually printed at the same time as the card is personalized, and again synchronicity must be checked after the cards are mounted in the mailers.

It is critical to track all failures at this stage of testing, as cards rejected because of a visual defect or synchronicity check may be completely functional. Development departments often seek to use such cards for testing, but this should be avoided at all costs!

As with the previous stages, a personalization number and unique reference is added, and once the personalization process is complete, the personalization lock is blown.

The data loaded onto the card (not including private keys) should usually be stored securely for a short time after the card has been personalized. This enables audit and recovery in case of problems. In some cases, data may need to be transmitted back to the host system. In any event, the card number, card, and batch status information should be returned for control purposes.

Data protection

Data-protection requirements must be taken into account during card personalization. We must first decide whether any of the data are regarded as personal, and if so whether the cardholder's permission has been obtained to store it on a computer system. According to the relevant national legislation, it may be necessary to provide a way of giving cardholders access to personal data stored on the card or to protect certain fields according to the wishes of the cardholder.

Electrostatic discharge and interference

Throughout the manufacture and personalization stages, and indeed throughout the life of the card, we must remember that a smart card contains a semiconductor device. It is therefore susceptible to problems such as *electrostatic discharge* (ESD) and interference from other devices. Equipment that was designed to handle traditional magnetic stripe cards may not prevent, and may even cause, such problems, and its suitability must be reassessed.

Issue

Cards may be issued to customers by a variety of methods, such as over a counter, through dispensing machines, and by mail.

In all cases, a control should be in place to register the card's change in status. Sometimes a barcode or other mechanism is used to identify the card at the time of issue; it need not necessarily be performed through the chip itself. In the case of postal dispatch, we know that the card has been sent, but not that it has been received; it is sometimes valuable to recognize these two different states.

Organizations that issue cards will already be familiar with the common forms of postal interception (e.g., accommodation addresses, false notification of changes of address) and the methods that can be used to defeat them. Two very common methods are to have a time gap between the card being issued and the beginning of its validity (during which time the issuer checks that the card has been received), or to require a separate action that includes an identity check (a telephone call, a visit to a branch, or use in an online terminal) to activate the card.

Smart cards can sometime be damaged in postal distribution systems: In particular, high-speed sorting machines often cause mail to turn a sharp bend when it hits the right pocket. Mailers should be designed to provide some form of protection, and where possible the card should be mounted so that the chip faces away from the front of the envelope to minimize damage from this source.

Loading/validation

Many types of cards (in particular electronic purses and stored-value cards) must be loaded before they can be used. This may be done at an ATM or other online terminal, at an offline *card validation station,* or by telephone.

In all cases, we should recognize that this is a standard double-entry bookkeeping procedure: A credit entry on the card account is matched by a debit to cash or to the account from which the money was drawn. As we will see, some purse schemes hold a *shadow account* on the central system, which should match the account held on the card.

The procedures for loading a stored-value card or purse usually involve higher security than when the card is used. Cards tend to be loaded with one large amount, which is then used in many transactions. Often a PIN or other password will be required for loading, even when the card may be used freely without a PIN. For example, customers with reloadable telephone cards are likely to be asked for a PIN when they press the special button that transfers a fixed amount of money from the account specified on the card or customer record to the card.

Use

As the card is used, we need to address the issues of logging and audit, card and cardholder authentication, and recovery from errors.

Logging

To maintain the complete record of the card's status, the first use of a card is often regarded as significant; this tells us that a transaction log for this card must now be opened and may give an opportunity to carry out extra checks to ensure that the card is being used by the person to whom it was issued.

When a card is used only as an identifier, it is unlikely that any data will change, and in these cases the use of the card need not be logged. Actions performed with the card may be logged for other reasons: For example, in an access control application, we may record the door being opened or a person arriving at work.

Where the operation results in data being changed, however, we usually record the change. If we are dealing with a transaction that will be logged in full somewhere else in the system, then it may be enough to increment a transaction counter within the card. This allows each transaction to be identified and provides a check that the record of transactions is complete. If the central transaction record does not exist or may be incomplete, the full transaction details should probably be stored on the card. Sometimes only the last five or ten transactions are stored in a rolling buffer, but for maximum security and audit, or if it may be necessary to deal with customer inquiries, a complete record should be held on the card. This may increase the memory requirements within the chip considerably.

Card and cardholder authentication

Whenever a card is used, the system or terminal should invoke whatever card authentication checks have been used—examples of these will be discussed in Chapters 11 to 16. This allows the card to identify itself positively to the terminal, and it may also be necessary to have the card authenticate the terminal (to prevent bogus terminals being used to extract data from cards). If there is a cardholder verification method (a biometric, password, or PIN check) on the card, this may be required for some operations but not for others. Thus, the decision to invoke this check can only be taken after the application has been selected, and it is usually part of the application logic.

Error recovery

The design of the transaction itself must take into account the need to recover from errors during the transaction, particularly where the transaction is not completed or where the card loses power during the transaction. This is common for contactless cards, but can also happen for a contact card if the cardholder is impatient and withdraws the card too quickly. In these cases, we usually want to cancel the transaction and return the card to its previous status, but we cannot do this if data has already been changed.

There are two ways of dealing with this. We can either change the data as we go along, but ensure that we are also building a log of the transaction so that we can undo it by reversing the changes. Or, we can employ the *two-stage commit* method used by larger computer systems, in which any changes are only made to a temporary copy of the data and are not committed to the main database until the transaction is complete. The first method is adequate for simpler applications, but for more complex systems, and particularly for functions that affect more than one application or data file on the card, the second method is considerably more secure and makes recovery easier.

Lost, stolen, and misused cards

Issues

When a card is lost or stolen, the outcome will depend greatly on the type of application loaded on the card. First we must be able to detect a lost or stolen card. In most cases we can prevent a thief from using a stolen card, and sometimes it will be possible to restore any value stored on the card to the genuine customer. In any event, we need to able to reissue a card to the customer.

For anonymous cards (such as disposable telephone cards) none of these is likely to be possible—when a card is lost, the finder gains its value.

First it may be necessary to verify that the card really has been lost: There are situations in which a user may benefit from reporting a card lost or stolen (for example, to deny responsibility for subsequent transactions on the card). A police report number or other proof may be required. The genuine user may also misuse the card, breach the conditions of the cardholding contract, or have the card withdrawn because of resignation. In all these cases we need to invalidate the card and prevent its use.

The situation becomes more complicated in the case of multifunction cards: The card may still be valid for other applications when one is blocked. Even if the card issuer contract is broken, the card may still contain value from another application issuer. So we need to be able to distinguish between actions that block the entire card and those that block only one application.

Detection

Ideally, the cardholder notices as soon as the card is lost or stolen and reports it to the scheme operator. The card is then put on a hotlist, and any online transaction will immediately be detected and stopped. In many systems, the hotlist is distributed to all terminals, immediately or at regular intervals; the terminal can then take action to block the card or can perform an online transaction to allow the scheme operator to take appropriate action.

If the cardholder does not report the loss, or in the case of misuse, the scheme operator must analyze incoming transactions to spot unusual patterns. For example, access-control systems look for consecutive uses too close together, while bank-card systems look for a rapid succession of transactions or a change in customer behavior. Some credit-card issuers employ neural networks to spot high-risk transactions and enforce a further check.

Block and unblock

If a stolen or misused card is detected, the terminal or host computer system can set flags that will block either the application or the entire card. It is fairly easy to reset a blocked application when the problem is sorted out (this might therefore be used in the case where a customer exceeds a credit limit); unblocking the card usually requires a further level of security because it affects all the applications on the card.

Financial cards are often blocked by sending the card a sequence of wrong PINs, until the card refuses to accept any more attempts. The corresponding method of unblocking is for the card issuer to give the cardholder a *PIN unblocking key* (PUK). This key (which has previously been stored in the card) allows the PIN try counter to be reset to zero only once; once a PUK has been used, a new one must be generated and stored in the card.

When a card or application is blocked, the card status must be changed accordingly, and usually an attempt will be made to recover the card or rectify the problem. Blocking procedures should be designed in such a way that they can be carried out at any terminal. Unblocking must be a more tightly controlled operation and is often only possible at specific terminals or after entry of a further security code by an employee of the scheme operator.

Reissue

Once a card has been reported lost or stolen (whether or not it has been blocked or recovered), there must be a procedure for issuing a replacement card. In some cases, this will be identical to the procedure for initial issue, but in other cases we must make provision for restoring the value on the card to the customer or application issuer.

For schemes with full *shadow accounting* (see Chapter 13), this will be an easy task, although often we must allow a sufficient period to elapse for all outstanding transactions to have been registered. In the absence of shadow accounts, it is virtually impossible.

End of life

Expiration

Cards should be given an expiration date. It may be possible to extend this date online or to ignore it in use, but the existence of a date is a final form of protection against cards that have defeated all forms of misuse detection. It also provides a natural route for upgrading cards and applications as technology and services move on. Bank cards are usually given expiration dates up to three years from issue. The cards themselves are usually expected to last longer than this, but some cards will already be showing signs of unreliability by this stage. In addition, it is difficult to extend the card's validity where the expiration date is printed or embossed on the card.

With multiapplication cards, we must ensure that no application has a longer life than either the card itself or any master application. Ideally they should all terminate on the same date. This is not always easy to organize.

Dispose or recover

Even an expired card may contain secret keys or algorithms which may be of use to a hacker. A stack of expired cards allows a test to be repeated many times or carried out in parallel, whereas few hackers will risk experiments on their own or on their friends' valid cards! It is therefore wise for a scheme operator to attempt to recover or have destroyed any cards that expire or fail.

For a closed user group such as a staff card, this should not be difficult. For other user groups, it may be necessary to arrange an incentive scheme, to issue cards only on an exchange basis, or to have a special transaction that validates the new card in return for capturing the old one. If none of these is possible, serious consideration should be given to the life of any keys used in the card; it may be necessary to update master keys and private keys at least as fast as the card expiration cycle.

Recycle

Card materials, particularly the PVC used in bank cards, are environmentally damaging, but recovered cards are often still fully functional. In some schemes it will be possible to reuse cards with very little effort, but in the majority of cases the card must be destroyed. Most manufacturers now have schemes for recycling a high proportion of card materials.

Part 3

Applications

11

Telephony and Telecommunications Applications

IN THE NEXT FEW chapters we consider some of the most important applications in which smart cards are used. For each application, we consider the security requirements and major issues, and how smart cards are or may be used to address these requirements. In this chapter we look at the smart-card applications that account for the majority of cards in use today: those in telephony, telecommunications, and broadcasting.

Most types of telephone systems suffer from some fraud and losses. True figures are difficult to establish, but the total fraud from public card telephones and analog cellular systems in Europe is probably of the order of $1 billion a year. In North America it is reckoned to be twice that figure. The growth of subscription television has provided another opportunity for fraud and subscription evasion, this time at the expense of the broadcasters and service providers. Although much smaller than telephone fraud, some companies estimate their losses at up to $20 million a year.

In computer networks, the problem is less one of direct fraud and theft than of confidentiality and integrity of data. The nature of the problem varies according to the type of network, with wide-area networks, and the Internet in particular, causing most of the difficulties. The expansion of this type of network in the 1990s has not always been accompanied by appropriate controls.

Prepaid telephone cards

Requirements

Public telephones using cash are expensive to build (they must be very robust to protect the cash from theft), expensive to operate (because of the need for cash collection), and unreliable (cash mechanisms become full, jammed, or vandalized). For many years now public telephone operators have exploited various forms of cards or tokens to overcome these problems.

Card telephones should have a minimum number of moving parts and must be able to operate reliably in a wide range of environmental conditions. The cards themselves must be easy to handle for all types of users, and they must be more costly to counterfeit than the maximum value on the card (typically $20 or so).

Although various forms of magnetic and optical cards have been used with success over the years, most telephone operators are now moving towards smart cards as the most effective card form (see Figure 11.1).

Standards

The so-called first generation of telephone cards are simple memory cards with an issuer identification area, logic that prevents the value on the card from being increased, and usually a relatively small maximum count (100–200 bits). The more up-to-date cards (known as the second generation) have maximum counts of 20,000 bits or more, card authentication mechanisms that protect against counterfeiting and

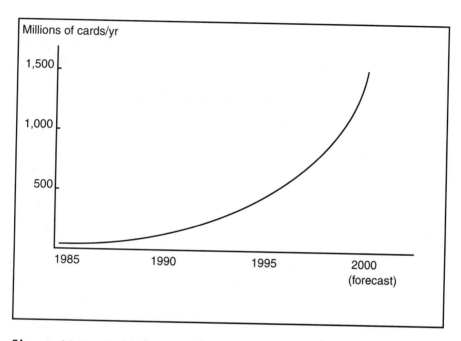

Figure 11.1 Market for smart telephone cards in Europe.

emulation, and logic that protects the card in case it is removed during a transaction; many also have a user memory area allowing number storage.

There are two common European standards for telephone cards:

- *EN1038* (see the appendix): Although this complies with the physical and electrical characteristics of ISO 7816, the data structures and operations are different. EN1038 was defined by an ISO working group that is in practice the TE9 working group of the European Telecommunications Standards Institute (ETSI).

- *T2G:* This is derived from the French telephone system and is a de facto standard by virtue of its longevity and the large volumes produced.

Issues

Telephone cards are often sold in small denominations ($2 or even less). Even the simplest smart cards cost 30–40 cents in large volumes, and an effective second-generation card currently costs around 60 cents. Although this cost will come down gradually, production volumes for these products are already around 400 million a year, and there will always be a need to increase security. So it is in the interest of telecommunication operators to maximize the value on the card and to add to the functions available on the card.

Both cards and telephones must have extremely high reliability, even though they are often misused. To avoid accidental misuse, the cards must be very easy and instinctive to use; it must, for example, be very clear when a card is correctly in position. The cards must also be able to handle such problems as being withdrawn at any time during a call; the status of the card at all stages in the transaction must therefore be recorded.

To minimize the overhead on the lines, card telephones must normally operate offline during a call and only exchange control information with a host computer on a periodic basis (e.g., once a day). Some modern systems are able to remove this limitation, at least partially, by using digital lines and distributed control systems. Nevertheless, the overhead involved in managing each call must be kept to a minimum. Individual details of cards used are not normally regarded as essential information for billing or operations.

When individual telephone cards are stolen, there is little the operator can do: The value is lost to the owner and gained by the thief. When large quantities of cards are stolen, however, they can be blocked by hotlisting if the card numbers have been traced throughout the distribution chain.

Several telephone operators in Europe use identical second-generation cards, although each carries the keys of the operator that issued it. This opens up the possibility of the cards being used across borders; each operator can make available its public keys to the others, thus allowing them to authenticate all cards. There are, however, contractual and regulatory issues involved here, as this is arguably a foreign exchange operation. Unit values may not be the same in each country. However, with the large count values available in the newer cards, this can also be resolved by deducting an appropriate number of bits per unit.

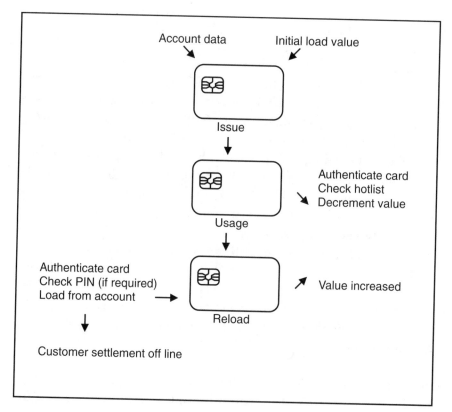

Figure 11.2 Reloadable card.

Reloadable and account cards

To avoid the waste of telephone cards, the cards can be made reloadable (see Figure 11.2). Value can be added to a card either using cash (in a controlled environment such as an office or shop) or by transferring money from an account.

A subscriber who already has a telephone account may have his or her account details stored on a smart card; when the card runs out of telephone units, pressing a special button on the telephone loads a fixed value to the card and charges it to the telephone account. The transaction is stored by the telephone and transmitted to the host computer at the end of the day for processing.

A less efficient option for the telephone operator is to charge every call directly to the account. This requires a record of every call to be maintained, along with the account details.

Both methods require the use of card authentication and hotlist checking. It is also usual to apply limits to the use of a card; there may be a maximum usage rate at one telephone within a period, or the limits can be stored on the card together with a daily usage counter.

Again, such cards can, in principle, be used internationally, subject to a commercial agreement between the operators. They can even be used in other applications; for example, one European telephone operator has devised a scheme in which a reloadable telephone card may also be used for public transportation in a number of cities. A separate store of value is maintained for each application or application issuer.

This comes close to the functionality of a full electronic purse, and indeed the CEN standard for electronic purses is based on work carried out by ETSI rather than any banking standard. Nevertheless, the scene is set for a conflict between banks and telecommunication companies as to who can issue electronic purses.

Conventional private exchange systems can also suffer from fraud. Many modern systems include facilities for direct inward system access, whereby employees can dial in from outside and not only access the internal network but make outbound calls at the company's expense. This kind of fraud is easier because many companies rarely or never change the passwords for inwards access. A smart-card token, used in conjunction with a public telephone or other smart-card-equipped telephone, not only provides an extra hurdle for the thief but can also enforce correct password discipline.

GSM telephones

Cellular telephony was introduced to most European countries at the beginning of the 1980s and rapidly proved very popular. The analog systems were developed by several companies independently, and each national implementation was different. Thus, telephones could not roam from one country to another (except in Scandinavia, where the networks were coordinated).

The original systems used analog speech delivery with unencrypted digital control protocols. As the networks became more popular, a number of possibilities for eavesdropping on traffic and for making calls on other people's accounts became known; this resulted in significant losses for both the telephone operators and their customers.

When the design for a Europewide mobile telephone system was being discussed and agreed, therefore, security was a high priority. The *global system for mobile telephony* (GSM) has now been adopted in many countries outside Europe, although there are competing standards in North America and in Japan. The GSM market currently accounts for 30 million cards a year and is forecast to rise to 100 million by the year 2000.

GSM security aims to:

- Authenticate the user (rather than the telephone);
- Protect the integrity and reliability of calls;
- Protect the confidentiality of calls and call-related data.

It does this by using a smart card known as a *subscriber identity module* (SIM) in every GSM telephone (see Figure 11.3). SIMs are either supplied as modules or complete ISO standard cards. A subscriber may have more than one SIM (for different networks, for example) or may use one SIM in different telephones (where the form factor of the card permits).

Phase 1 of the GSM specification called for a card with 4 kb of E^2PROM to accommodate subscriber identity and keys, as well as private storage for frequently used numbers. Phase 2 (currently in use) calls for 8 kb, and a 16-kb standard is planned. The additional memory is used to store additional personal data rather than just telephone numbers. GSM defines its own operating system, which protects the data in the card so that it can only be accessed by authorized applications. Although the structure would permit applications on the card to be issued by different application issuers, no current schemes use this facility.

GSM cards support PIN checking, application and card blocking, and unblocking by means of a PUK (as described in Chapter 10).

Figure 11.3 GSM telephone with SIM. (Photograph courtesy of Motorola Ltd.)

Television decryption

Requirements

Satellite and cable television companies seek to obtain a proportion of their revenues from subscriptions, particularly for premium services, and *pay-per-view* income (although advertising remains the main source of revenue for most companies).

Subscription services are transmitted in an encrypted form, and the decryption is carried out in a *set-top box,* which is fitted with a smart card (see Figure 11.4). The cards are sold on a subscription basis and may also contain credits for pay-per-view watching.

The smart card also provides the means to control other aspects of viewing, such as programs suitable for children. Some industry sources also believe that it will allow selective downloading of advertising or messaging or even emergency messages broadcast to specific areas or groups. Although satellite signals reach all countries, governments may control the sale of cards for specific services thought to be undesirable.

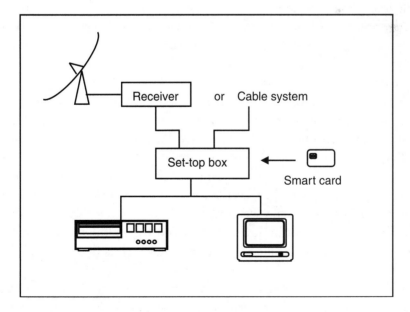

Figure 11.4   Television decryption.

Weaknesses and responses

The Videocrypt encryption method and keys used in analog satellite television have been repeatedly cracked; software containing the algorithms and codes has been published on the Internet, and pirate cards are sold openly in several countries, notably in Germany. A North American pirate card ring raided in mid 1996 was estimated to have produced over 30,000 cards. Figures of $20 million a year quoted for lost revenue to the networks are probably exaggerated, as most of those who bought the pirate cards would not have paid the full price for the service. The cost of replacing the smart cards was much less than the cost would have been if the whole set-top box had needed replacement.

Nevertheless, this is the most serious threat to the credibility of smart cards as an aid to security, and it is important to know what lessons have been learned.

Much of the problem can be traced to an underlying weakness in the encryption method. Modern encryption relies on digital techniques and is applied to digital data. When applying it to analog television signals, the original designers simply divided the signal into finite slices. Each television line is divided into two parts, and the two parts are rotated. Every few seconds a new key is transmitted in an encrypted form in the vertical blanking interval (the gap between pictures usually used by teletext and overlay signals); this is decoded using the master key held in a smart card in the set-top box.

There is therefore a strong correlation between the encrypted signals and the original; it is even possible, with a high-speed signal processor, to match elements of successive lines so that the signal can be decrypted without use of the algorithm at all! For the hacker, however, the availability of large quantities of matching plaintext and ciphertext reduces the key search time to very low levels once the algorithm is known. The second level of encryption involved in deriving the master keys can also be simplified by building a unit that appears to the card like the set-top box, but which is directly controlled from a PC.

These cards were replaced in 1994 with cards using two-way authentication, a much stronger algorithm and longer keys; since then there have been reports of hacks, but open sales of pirate cards have stopped. Satellite broadcasters are nonetheless keen to move on to digital broadcasting, where the full range of modern encryption techniques can be used.

Computer networks

Computer system access

Most computers are accessed by a combination of a user identification (user ID) and password. Enforcing good password discipline is notoriously difficult; the same controls that make it difficult for outsiders to guess passwords also make the passwords difficult to remember. Even with good discipline, valid passwords can be obtained by looking over people's shoulders, by monitoring data traffic, or in some cases by hacking into system files.

A much more secure form of control involves the use of an intelligent token in conjunction with a password. Now that smart-card readers are becoming easily available for PCs, they are one of the easiest forms of tokens to implement. The password (or a PIN number) is checked by the card and need not be stored anywhere on the system. When the card is inserted, the system asks for the user ID; this may be given by the card or by the user. The system then authenticates the card (using a zero-knowledge or similar test, as described in Chapter 5), and the card authenticates the user. The use of random or time-dependent functions is important to avoid replay attacks.

The card may also contain a user profile, including preferences and membership of groups (where access to specific system functions is specified by groups). The access rights themselves should normally be held within the system and associated with the data files or applications in question.

When the card is removed, the user should be logged off.

Several companies now offer software to perform access control and preference setting in this way. Further information about protecting computer system access using smart cards and similar technology is given in other publications. (See the bibliography.)

Confidentiality of data and programs

Systems for protecting stored data must always strike a balance between confidentiality and ease of use. If a system is designed to protect against every possible attack, the smallest malfunction can render the data unreadable. Storage of keys is a particular problem and is an area where smart cards can be of assistance. They also help to enforce certain specific types of access control, notably where individuals are held responsible for authorizing access to data. In these cases, the keys are held on a smart card, which is kept by the person responsible. When designing such a scheme, it is important to set out the security objectives clearly and to acknowledge, where trade-offs must be made, what decision has been taken.

Smart cards are also used for storage and loading of keys into encrypting modems and other hardware devices used for encrypting data during transmission from one system to another. The procedures described by the manufacturers of these devices for setting up and maintaining these systems must be carefully followed if their security is to be maintained.

The Internet

The wide scope and anarchic nature of the Internet give rise to several special issues. Smart cards themselves cannot enhance the security of the Net, but they are used in two important applications.

First we should distinguish between the Internet as a means of access and the content to which it gives access (mainly email and pages on the World Wide Web).

Internet and Web access points

The protocol used for passing messages between computer systems on the Internet is called transmission control protocol/Internet protocol (TCP/IP). TCP is concerned with message handling and IP with addressing. TCP/IP was designed to allow messages to be routed through the network using any available connection. There is no protection against any of the intermediate computers reading, altering, or destroying the message. The next version of the Internet protocol (IP version 6) will include a moderate level of protection but will not be widely implemented for several years.

Systems connected to the Internet are more likely to be hacked than those on private networks or other wide-area networks. Companies can protect against this to a large extent by using *firewalls*, which only allow certain types of traffic and users to enter their systems. But firewalls do not prevent eavesdropping (other systems gaining access to data transmitted) unless the data is encrypted. And few firewalls are able to prevent files from being received if they are "invited" in. These may include *cookies* (programs that store data about the user and transmit that data to a host when the host computer is next accessed) and other nastier snooping programs (which intercept data and store it for later transmission).

Data content

Another characteristic of the Net is freedom of information: There is a wealth of data on the Net about encryption products and ways of breaking them. Newsgroups such as *comp.security* and *alt.security* are essential reading for designers of secure systems. They help to give an understanding of the mentality of hacking as well as the techniques used. The U.S. government has been particularly concerned about the possibility of strong encryption techniques becoming available through the Internet and limits the length of asymmetric key systems used on the Net.

Internet mail

Probably the most widespread application on the Internet is electronic mail. Much email is not confidential, and those sending and receiving it can risk the small chance that someone else reads or even alters it. In certain cases, however, a higher level of privacy or integrity must be ensured. This applies most often to commercial data such as orders, stock lists, and communications with field staff, and to subjects that are covered by data protection or similar legal restrictions (such as communications between doctors about a patient).

There are several standards for increasing the privacy of email. Most of these include authentication of the originator and recipient, encryption of the data, and integrity checking of the whole message. They typically make use of a public key system for authentication and transmission of a symmetric key, which is then used for encrypting the body of the message. The systems most commonly used are Privacy Enhanced Mail (PEM), Riordan's Internet Privacy Enhanced Mail (RIPEM), and Pretty Good Privacy (PGP). Most email software systems will support one of these; different versions must sometimes be used within and outside the United States for export control reasons.

The private key for authentication may be stored most conveniently on a smart card; it is then relatively secure and can be used on any computer fitted with a smart-card reader. Some packages will now support this option. The public key can, of course, be published in an email message or on a Web page.

Internet purchases

The commercial side of the Internet has grown rapidly since it was made available in 1994. Internet shopping is seen by many as a key factor in the future of retailing, but in practice, the growth of Internet commerce has been slow. Many companies have come to realize that selling is only part of the story; delivering the goods usually requires traditional mail-order skills. The international nature of the Internet means that companies have to be prepared to do business all around the world, under a wide variety of unfamiliar legal systems.

Another barrier is the difficulty of establishing trust. A flashy Web site can be a front for a 15-year-old in a bedroom or a bogus company aiming to defraud potential purchasers. Even well-established companies may have difficulty demonstrating their validity on the Net.

Even when the purchaser accepts that he or she is buying from someone safe and recognizable, making payment is difficult. Developers often forget that the majority of would-be customers do not have a credit card. Those who do are often not prepared to send their card numbers over the Net, with its acknowledged insecurity.

The difficulty of making secure payments over the Internet has been widely recognized and seen as an opportunity by many companies. As a result, there were over 60 different payment systems being offered on the Internet by the end of 1996. When making a payment for goods or services on the Net, merchants may offer all or any of the following methods:

- *Open payments:* Card numbers and expiration dates can be sent to merchants openly; many people are surprisingly happy to do this, particularly for smaller amounts. The credit-card issuers do not encourage this form of transaction; they permit it but do not offer the usual guarantee of payment.

- *Secured link:* Most popular browsers have a secure mode, which uses public-key cryptography (with quite a short key) to authenticate the parties and encrypt the transmitted data. The latest versions of software allow the browser to read the authentication keys from a smart card, and in the future this could be a function of a smart credit, debit, or purse card.

- *Trusted third party (TTP):* Several organizations have set themselves up as intermediaries; both merchants and customers must register with the TTP. Customers usually register a credit card or bank account number and continuous payment authority. The registration process may be carried out offline (by telephone, mail, or fax). The TTP allocates each party an account number, and is responsible for collecting funds from customers and crediting merchants. Only the account number is transmitted online, and this may be checked against a delivery address by the TTP.

- *Digital cash:* Again normally operated by a third party, digital cash consists of software certificates which are in effect virtual banknotes, with a fixed value. A customer must purchase these certificates in advance from the scheme operator. When a payment is to be made, the customer's software issues the appropriate number of certificates to the merchant. The merchant returns the certificates to the operator, which checks them and credits the merchant's account.

- *Electronic purse:* These products will be discussed in more detail in Chapter 12. A payment can, however, be made directly from an electronic purse; the security inherent in such schemes ensures the integrity and confidentiality of the transaction. In most cases, it also ensures that the merchant is registered with a valid card-payment scheme.

The companies offering these services are generally not banks. They are information companies, and only some of them use conventional banking services (such as the credit-card networks) to clear transactions. The banks see this as a clear threat, but have yet to develop a consistent counterapproach. The banking industry's approach to electronic transactions will be described in Chapter 12.

12

Financial Applications

I N THIS CHAPTER we look at the ways smart cards are used in financial and banking environments and, in particular, at the opportunities for chips to replace magnetic stripes on payment cards.

Bank cards

Functions

We are so accustomed to seeing and using bank cards that it is sometimes difficult to distinguish between them. Traditionally, banks have issued several distinct types of cards:

- *Account cards:* These serve as an identifier only and are typically issued for accounts where no other type of card would be appropriate (e.g., savings and notice accounts). They avoid the need for customers to remember account numbers or for bank staff to search lists.

- *ATM cards:* These are commonly issued to customers who have no credit, including children and new customers. They may only be used in ATMs connected (directly or through an inter-bank link) to their home bank network, which means that the transaction is always checked against the current balance on the bank's own computer system.

- *Debit cards:* These are now the most common cards in Europe. They can be used in ATMs and in retail shops, and they are always linked to a current account. Transactions are posted to the account as soon as they are received; in the case of retail transactions, this may be immediate, the next day, or after processing of a paper voucher. Some debit-card schemes (mostly national schemes) insist on electronic processing, but the transaction is often authorized by another bank or by the card scheme (Visa, Europay, or Mastercard) rather than by the card issuer; this is called *stand-in processing.*

- *Credit cards,* the oldest form of bank payment cards, allow a customer to pay on credit, up to a preset limit. Credit cards may be used online or offline, and the level of authorization by the issuer is lower than for debit cards. The risk of both credit losses and fraud is therefore greater; thus, the charges to both merchants and cardholders are higher than with debit cards.

- *Charge cards* (also called *travel and entertainment cards*) are mostly issued by nonbanks; they have no explicit spending limits and work on a deferred payment basis rather than rolling credit (the balance must normally be paid off in full every month). Retailer charges are highest for these cards, reflecting not only marketing costs but also the increased risk of default.

Nearly all debit, credit, and charge cards work within the framework of a payment scheme (such as Mastercard or American Express). The schemes set up relationships between merchants and their banks (merchant acquirers), cardholders and their banks (card issuers), and acquirers and issuers (see Figure 12.1). Merchants normally agree to accept cards from all issuers. Because issuers and acquirers upgrade their systems and cards at different rates, it must be possible for the newest card issued by

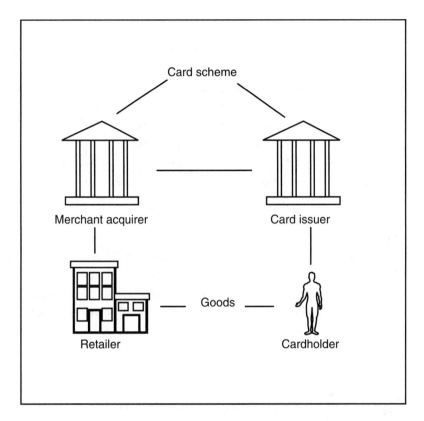

Figure 12.1 Standard model for card payments.

a large bank in, say, Germany to be accepted and processed by a small retailer or hotel in Africa or Asia. The need to maintain backwards compatibility is one of the biggest limitations on the speed of technology diffusion in the card schemes.

For some years now, the major card schemes have also maintained a high level of *interoperability*. Cards from all manufacturers and issuers can be accepted by a single terminal at the point of sale. This requires the use of standards, not only for cards but for the electronic messaging systems that transmit and process transactions. The card standards, as we have seen, are based on the ISO 7810 family, while for messaging most countries use a variant of one of the Visa or ISO 8583 standards.

The advent of the smart card has allowed banks to move on to the next stage: the *electronic purse*. While credit cards offer "pay later" and debit cards are "pay now," electronic purses are prepaid. The term *stored-value card* is also used, particularly when talking about a card which can only be used to pay for goods and services from one source (a vending-machine card, for example, or a multiple-ride ticket for a bus or train service). As soon as it becomes possible to buy goods and services from several different suppliers, this enters the preserve of the banks, and the card is then referred to as an electronic purse.

We will also start to see *electronic passbooks,* for accounts that traditionally have been maintained using printed records in a book. Most passbooks carry a magnetic stripe, which duplicates all or some of the printed data in the book, but these suffer from the same security weakness as other magnetic-stripe systems. The use of smart cards will allow the book to be used as a more secure identification of the account holder.

Banks are also seeking to hold more customer data on smart cards held by the customer. This is to offer an increased range of services offline and, in some cases, to overcome data protection limitations on data held online.

Attacks

The most common source of fraud on cards is theft. When a card is lost or stolen, the thief may be able to use it in an outlet where PINs are not required—particularly if the retailer or shop assistant can be relied upon not to inspect the signature too thoroughly. As soon as the theft is reported, the card may be hotlisted, but this will still only affect transactions in electronic terminals until the next paper hotlist is published. With over 20 million lost and stolen bank cards worldwide, even electronic lists must be prioritized before transmission.

When the PIN is stolen with the card, considerably greater damage can be done because the card can then be used at ATMs and PIN-enabled merchants.

Thieves can also obtain cards by intercepting them before they reach the valid cardholder. There are various techniques for doing this, but the effect is to obtain an unsigned card, so that the thief can sign it. The valid cardholder is unlikely to know that the card has been intercepted, so the time delay before the card is reported lost will probably be longer.

Other attacks on magnetic-stripe cards include the following.

- *Replay:* Merchants swipe the card several times or reproduce the card number or transaction electronically.
- *Counterfeit:* A card that appears to be valid is manufactured (it is often a copy of an existing card) and used for transactions.

PINs and other electronic checking of cardholder identity protect against several of these attacks, while others can be effectively prevented by the use of online transaction authorization. PINs with magnetic-stripe cards are only checked online, as they involve the use of a master key. In the United States, online authorization is used for almost all transactions, but in most other countries the cost and complexity of the networks involved means that a significant proportion of transactions must be carried out offline.

Credit/debit cards

Requirements

The prime requirement for a smart credit or debit card is to reduce the opportunities for fraud without increasing the need for online authorization. In fact, as banks seek to drive more and more transactions towards cards rather than the less profitable media of checks and cash, the threshold for transactions is moving downwards. As a result, card transactions must be made economic for transactions of a few cents or even less. This creates a firm requirement for offline transactions.

The first requirement is therefore to be able to authenticate the card offline. All cards are issued by an issuer that belongs to a scheme. All terminals linked to that scheme must as a minimum have loaded a key belonging to the scheme (usually its public key). This enables the card to be authenticated using a hierarchy: the card itself contains a certificate issued under the secret key of its card issuer, whose public key, signed by the scheme, also features on the card. This form of *static data authentication* (SDA) can be checked by any terminal loaded with the scheme public key and does not require any cryptographic processing by the card (see Figure 12.2).

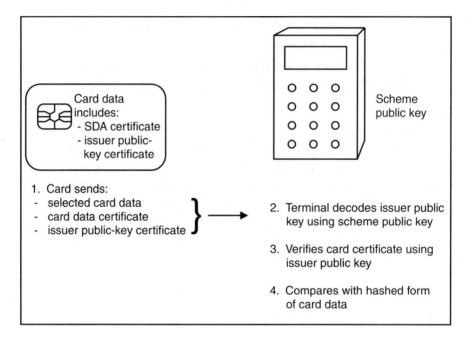

Card data includes:
- SDA certificate
- issuer public-key certificate

Scheme public key

1. Card sends:
- selected card data
- card data certificate
- issuer public-key certificate

2. Terminal decodes issuer public key using scheme public key

3. Verifies card certificate using issuer public key

4. Compares with hashed form of card data

Figure 12.2 Static data authentication.

It would not, however, detect a card that had been made using an exact copy of the original card data (including the certificate). This requires *dynamic data authentication* (DDA), in which the terminal issues a challenge to the card and the card responds with a signature based on not only the card data and its own secret key, but also the challenge. This method, which is another form of the zero-knowledge identification test described in Chapter 5, will detect anything other than an exact copy of the card itself. The card must, however, incorporate the public-key encryption functions and be powerful enough to carry these out without delaying the transaction.

A further check of the card authentication can be carried out if the transaction occurs online. In this case, a symmetric key can be used because the keys do not have to be distributed throughout the terminal network. Most bank cards can perform DES encryption within the card.

The next requirement is for verification of the identity of the person presenting the card. This is essential for any unattended environment, and even in retail shops it removes the need for assistants to check signatures or other subjective criteria. Retailers do not like having to perform these checks on behalf of the bank; challenging someone on the basis of their signature or appearance can provoke a confrontation. The range of biometric checks available on smart cards was described in Chapter 6, but the PIN remains the most common CVM by far in banking applications. Where PIN checking is used with smart cards, the PIN may be checked offline, by the card.

Last, the card issuer seeks to ensure that customers do not exceed their credit limits. With most types of accounts, there are ways of carrying out transactions other than using the card (such as making payments to the account). So the card cannot carry a full record of the current balance. The card is therefore used to force the transaction online under certain conditions; these would include high-value transactions, a long sequence of offline transactions, or a volume limit which can be customer specific. These limits can in principle be varied dynamically, allowing a very high level of risk management.

This model allows card operations to be extended not only to lower value transactions and to unattended terminals, but also to those countries where the telecommunications infrastructure is not sufficiently well developed or where telephone calls are too expensive to permit a high proportion of all transactions to be carried out online.

It also offers the possibility of *cardholder controls*. Some customers want to be able to control their spending very tightly and are prepared to put up with some inconvenience or delay to achieve this. For others, convenience and speed are more important than detailed records of all transactions. The smart card allows either of these requirements to be accommodated within a single scheme. Customers may also opt for different levels of control on different cards on the same account; this would allow a parent to give a card to a child, subject to strict limits on expenditure per transaction, expenditure within a day, or on the type of goods that may be bought.

Standards

The need to maintain global interoperability has slowed the development of standards for smart payment cards. The card-payment schemes are owned by their members, whose priorities cover a wide range. Four of the five main schemes are headquartered in the United States, where the high level of online authorization reduces the pressure for smart cards to be introduced to control fraud.

In France, however, the smart card became a national technology and source of pride in the 1980s at a time of high and rising fraud. The French banks had a single association (the *GIE Cartes Bancaires,* known as CB) which controlled card-payment standards. It was tasked with developing a chip-based system, which started trials in 1987. In 1990 the decision was taken to proceed with a rollout and since 1993 all domestic cards have been smart cards.

This program has been exceptionally successful; the level of fraud on French-issued cards is around 0.025% of turnover—less than a quarter of the worldwide average and little more than a tenth of the level at the start of the program. Much of the residual fraud takes place overseas. And the level of online authorizations, at around 10%, is also very low by world standards.

Because this scheme was put in place before any other countries had considered adopting smart bank cards, the standards for the French program were created by CB and reflect the structure of the French market and network. The chip was placed close to the top left-hand corner of the card, in a position which overlapped the magnetic stripe on the back of the card. This is a potential source of damage to both magnetic-stripe card readers and chips; thus, when the international standard ISO 7816 was written, the chip was moved to the new position described in Chapter 8. French cards will start to migrate from the Carte Bancaire "high" position to the "standard" positioning of the chip during 1997.

The CB application also reflects that it is a national scheme with a high level of coordination between issuers and that all the cards concerned are debit cards. When the major card schemes got together to design an application, their starting point was quite different. They acknowledged the need for the card-authentication method to be largely independent of the terminal, for issuers to vary the combination of applications and cardholder verification methods on the card in any way they chose, and for new applications to be introduced without affecting the existing card or terminal base.

The Europay–Mastercard–Visa (EMV) standards (see the appendix) therefore define the card and card interface in considerable detail, but many of the functions they describe are optional and many variables are in the hands of the card issuer. EMV provides a common language and a basis for interoperability between schemes but does not of itself guarantee that interoperability. Each of the card schemes has set up a smart-card payment system using EMV.

The EMV standards were first published in June 1994 and were added to and updated in 1995 and 1996. They draw heavily on the corresponding ISO standards. However, they put these standards into the context of a financial card, and they specify certain common functions, such as static data authentication, application blocking and card risk management.

EMV was defined with credit and debit cards in mind, but with some additional functions an EMV-compatible card could also contain other applications, such as an electronic purse. Although there can be several applications on a card, EMV in its present form does not define any mechanisms for communication or protection between applications—it could be regarded as a single-tasking operating system.

The EMV mechanisms for card authentication include static data authentication, dynamic data authentication (both using public key encryption as described earlier), and online authentication (using triple-DES with a derived key). Message authentication is also DES based, and there is provision for PINs, oncard biometrics, or signature checking.

The general model of an EMV transaction is described in Figure 12.3.

The French banks will start to migrate their cards to an EMV platform during 1997, and this standard can be expected to form the basis for any future smart credit/debit cards.

Procedures

There are few changes in the procedure when a card is used. The card must be inserted into the correct terminal to read the chip (data on the magnetic stripe identifies the presence of a chip). A PIN or biometric may be checked—the PIN entry procedure must be controlled so that only the cardholder sees the number, and terminals themselves must not offer the opportunity to read the PIN electronically as it is input.

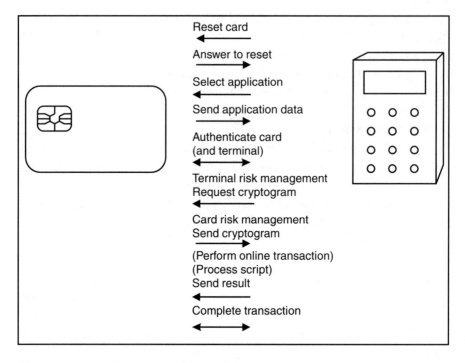

Reset card

Answer to reset

Select application

Send application data

Authenticate card
(and terminal)

Terminal risk management
Request cryptogram

Card risk management
Send cryptogram

(Perform online transaction)
(Process script)
Send result

Complete transaction

Figure 12.3   EMV transaction model.

Smart-card issuers have available a large amount of extra data associated with each transaction. Some of this data may be useful in spotting patterns that might indicate fraud, and the systems used for detecting these patterns on magnetic-stripe systems must be updated. Issuers must also make decisions as to the meaning of different failure conditions—does it mean that the card has been tampered with or could an item of equipment be faulty?

The biggest changes for card issuers, however, are in the areas of key management and risk management. Although all banking standards require the use of host security modules for the generation, storage, and processing of keys, the manual procedures and controls are still important, as we emphasized in Chapter 10. Every financial institution has a data-security officer, a data-protection officer, and a risk-management department, and these people will all be closely involved in the implementation of a chip-based credit- or debit-card scheme.

Although larger issuers may personalize their own cards, most banks subcontract the personalization process to the card manufacturer. This means that the manufacturer's processes must be bound in to the security and audit processes of the bank. Since a manufacturer will work for several banks, this requires an exceptionally tight control of procedure, data, and in particular keys.

Electronic purses

Requirements

Electronic purses are usually associated with smaller purchases where there is less need for individual controls on each transaction. The procedure for using an electronic purse is usually kept very simple (see Figure 12.4), but this simplicity conceals a process with an exceptionally high need for security.

Electronic purses have a special position because of their relationship with cash. If an electronic purse can be counterfeit or the amount on the purse manipulated, this is the same as forging banknotes. Central banks are also concerned about the monetary control aspects of electronic purses and will not allow any commercial business to create tradable value denominated in a national currency.

The Working Group on EU Payment Systems, in a report for the European Monetary Institute [1], concluded that electronic purses, where the card issuer and the provider of goods or services were separate organizations, should only be issued by credit institutions regulated by a central bank.

The Bank for International Settlements also considered electronic purses in a report for the G-10 central banks [2] on the security of electronic money. It concluded that purses must use very long keys (128 bits for DES and 2,048 for RSA) to be considered secure for many years to come. The Bank also recognized the importance of overall risk management and independent security checks in the design and management of such schemes.

Security is also important for the issuer, who is at risk if extra value or transactions are created. Merchants stand to lose money if transactions

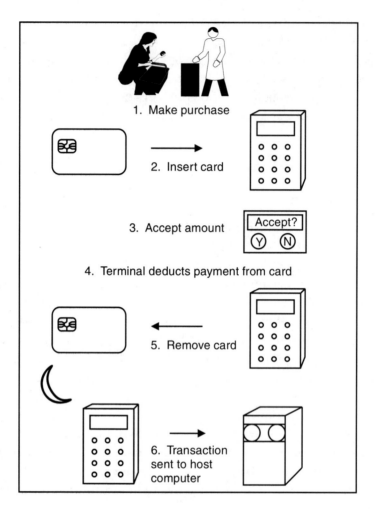

Figure 12.4 Using an electronic purse.

(which are usually only held in electronic form) are lost. Both of these are less concerned with the effect on the consumer if the card is lost, but consumers are less likely to hold significant value on the card if they feel that they may lose it all when they lose the card. Some schemes are able to block a lost or stolen card, calculate the value left on the card, and issue a replacement.

Cards are a convenient way to carry money without the need to have the correct coins for a given transaction. They are therefore most popular in unattended applications or on public transportation. These correspond with the situations in which there is no human operator to check the card or cardholder. The relationship of trust is no longer between the retailer and customer but between the card and terminal: the card issues the guarantee of payment, digitally signed by the card issuer.

Those who travel across borders have a far greater problem carrying coins and notes; they end up holding a significant float in different currencies or exchanging cash at a large loss on their return. Few electronic purse schemes can handle multiple currencies today, but the number that can is growing. With the approach of European monetary union, electronic purses that can operate in euros just as easily as in francs, pounds, or pesetas will become a valuable tool for easing the transition. There is a requirement for an application to convert currencies within the purse, using a common reference held in the terminal.

Another area where we can expect electronic purses to be used more widely is for micropayments. Today it is not worth the effort for a company to charge a fraction of a cent for, say, access to a piece of information. The transaction costs exceed its value. The information is therefore either provided free or not at all. If payments can be made electronically, at zero marginal cost, then it becomes worth allowing access to the information at a tiny charge. This principle can be applied to other forms of goods or service; retailers no longer need be limited to charging in whole pennies, for example. Electronic purses provide one possible means to store and exchange these small units of value and to transmit them to a host system in bulk.

Types

The simplest form of electronic purse is the *stored-value card*. We usually take this to mean a card with a single store of value, issued by a company and only usable to buy goods and services from that company. Stored value cards are used, for example, in vending machines, retail-loyalty schemes, and for cashless payment in clubs and canteens. Their function is almost identical to that of a telephone card, and they can usually make use of the authentication methods used by those cards.

When a card is designed to be used, like cash, in open circulation, then it is considered to be an electronic purse. Most electronic purse

Figure 12.5 Electronic purse balance reader. (Visa Viewer for Visa Cash program.)

schemes still work within the issuer–cardholder–retailer–acquirer framework shown in Figure 12.1. This means that transactions are asymmetric: the retailer terminal is in control of the card. Transactions are normally made by deducting value from the card, although some schemes will also allow refunds to the card. The issuer may make available a balance reader (see Figure 12.5) to allow checking of the current balance on the card.

The card issuer normally loads the purse with such a scheme via an ATM or specially adapted telephone. This gives the issuer regular opportunities to check and reset the flags and counters held in the card.

To be fully interchangeable with cash, however, cards must have the ability to make *card-to-card transactions* outside the payment schemes. This allows cardholders to make interpersonal payments to members of their own family, friends, or casual service providers who are not card-scheme merchants (e.g., domestic staff or window cleaners). This facility is only provided by one major card scheme today: the Mondex scheme devised by National Westminster Bank in the United Kingdom and now licensed by many banks around the world. Mondex offers a wallet (Figure 12.6) that can read the balance on the card in any of the five currencies stored,

Figure 12.6   Mondex wallet. (Photograph courtesy of Mondex International Ltd.)

allow the card to be locked or unlocked, and store Mondex value, which can be transferred to or from any Mondex card.

This feature of Mondex has been criticized by some bankers because it does not allow checking and audit of every transaction. Value can in principle circulate for long periods without being cleared through a bank. Although the incompleteness of the record does demand a higher level of security and, as we shall see, some additional checks, the comparison with cash shows that there is a demand for card-to-card payments and that these need not always be under control of a bank.

The simplest electronic purses are, like most telephone cards, disposable: They are loaded before issue and thereafter the value can only be decremented. The more elaborate schemes allow reloading of the purse (some schemes have both types of cards available). Cards can be reloaded at an ATM, through a specially adapted telephone, or using cash at a *card-validation station* (Figure 12.7).

Disposable cards are almost always anonymous; they are sold by appointed retailers or company service points and carry no link with the customer. For cards to be reloadable from an account, they must be personal to that customer. They will either carry details of the account to which they are linked or a reference number which can be linked to the account. In these cases, the card will probably be personalized by printing the customer's name on the card as well.

Many electronic purses are issued to customers who have no bank account or who would not qualify for another form of bank card. A personal electronic purse card is the first step towards establishing a banking relationship for many such customers and banks.

Even with personal cards, transactions can be made anonymously when a system such as Mondex is used. But other banks see this as a disadvantage because part of their aim is to know more about their customers' spending habits, to be able to match the products offered to customer requirements.

Cleared schemes maintain a full shadow account for each card and will check that the balance on the card is consistent with the balance on the shadow account every time the card connects online or a transaction is posted. The match may not be exact, because some transactions may not yet have cleared, but the card's transaction counter will say when this is the case.

Figure 12.7 Card-validation station. (Photograph courtesy of Datastripe Ltd.)

Uncleared schemes maintain security by the use of statistical measures, trend analysis, and exception monitoring. In addition to a log of the last few transactions, the card itself will contain an exception log which records details of unsuccessful transactions and which can be inspected by a bank terminal. The BIS report referred to earlier recommended limits on the volume or number of card-to-card transactions to maintain the flow of security monitoring information.

It would also be possible to devise an electronic purse based on the use of certificates representing finite units of value, as in the software schemes described in Chapter 11. There is a commercial link between the Mondex scheme and DigiCash, one of the major issuers of certificate-based electronic money. To date, however, all smart-card-based schemes use the principle of storing a single balance on the card.

Status

Although electronic purses have only been around since the beginning of the 1990s, there is already a wide variety of schemes in operation with differing characteristics. The longest established scheme is Danmønt, which started trials in 1992 and by 1995 was well established in most large towns in Denmark. Danmønt is an open, cleared system; it started using only disposable cards but introduced reloadable cards in 1996.

The Belgian Proton scheme was also an early development; it uses reloadable cards and one of its special features is a secure *payment module* common to every card accepting point. Like most of the other schemes listed in Table 12.1, Proton is a single-currency scheme, but it can be adapted to work under a multicurrency "umbrella." It has been licensed to many other countries including Brazil, Switzerland, Sweden, and the Netherlands.

Visa has promoted several schemes under its Visa Cash brand. Not all of these are compatible, but the standards are evolving and an international Visa Cash standard will be published during 1997. The U.K. pilot scheme in 1997 will be the first Visa Cash card to use dynamic data authentication, which is expected to become the norm for electronic purses and other cards intended primarily for offline use.

At this stage, most of the schemes described are self-contained and incompatible with one another. Although an international standard for

Table 12.1 Selected Electronic Purse Schemes Operational in 1997

Country	Scheme Name	Operator	Status (end 1996)	Features
Australia	Transcard	Card Technologies Austr.	Trials in Sydney; further cities imminent	Contactless cards
Austria	Quick	Europay Austria	Operational (3 million card)	Contactless cards
Belgium	Proton	Banksys	Being rolled out	Reloadable cards; terminals require SAM
Brazil	Proton	Banco do Brasil	Pilot	Proton model; includes salary payment
Canada	Mondex	RBC; CIBC	Pilot Guelph	Mondex model
Denmark	Danmont	Danmont	Fully operational since 1994	Disposable and reloadable cards
Finland	Avant	Bank association	Being rolled out	Disposable and reloadable cards
Germany	Geldkarte	ZKA	Field trials	Reloadable cards
	PayCard	Telekom	Trials	Reloadable cards; use in public transport and telephones
Italy	P-Card	Electronic Banking Sys.	Announced	Reloadable; retail loyalty functions
	Cassamat	Bank co-op	Operational but local	
	MINIpay	SSB (bank owned)	Trialling Turin	
Netherlands	Chip Knip	Banks	Rollout imminent	Proton based
	Chipper	PTT	Rollout imminent	
Portugal	PMB	SIBS	Fully operational	Reloadable cards; uses SAM

Table 12.1 (Continued)

Country	Scheme Name	Operator	Status (end 1996)	Features
Russia	Zolotaya Korona	CFT	Several regional schemes operating	Reloadable; multiple currencies
Singapore		NETS	Trial in progress	
Spain	Visa Cash	SEMP	12 independent schemes operational	Now branded as Visa Cash
Sweden	Proton variants	All major banks	Rollout imminent	Proton model
Switzerland		Telekurs	Rollout started	3m ec cards; Proton based
Taiwan		FISC (Government)	Operational	
Zambia	Meridien Card	Meridien Biao SA	Operational	Reloadable
U.K.	Mondex	Mondex UK	Piloting Swindon, Exeter, and York	Offline/card-to-card operations
	Diamond	Barclays; Lloyd; etc.	Pilot due 1997	Visa Cash
U.S.	Mcard	First of America	60,000 cards in University of Michigan campus	
World	Clip	Europay	Ready for pilots	
World	Visa Cash	Visa	Piloting Atlanta, Gold Coast, S Africa, Bormio	Based on Danmont technology

electronic purses exists, it originated in the telephone sector and does not conform to the banks' preferred architecture. In the medium term, it is likely that two parallel systems will emerge, but the independent schemes will still be in use for many years to come.

Online transactions

Transaction authorization

Any type of transaction, such as sending a purchase order, placing an advertisement, or making a payment, can be authorized using a smart card. A public-key system is used, with the secret key held on the card and the public key made available to the other party. A digest of the message is created by the software; this is signed by the card using the secret key and checked by the recipient using the public key.

If the process is repeated in reverse (when the transaction is acknowledged), then the originator is able to authenticate the recipient. This mechanism provides a complete check of the integrity of the message and neither party can subsequently repudiate the transaction. It is frequently used by financial institutions in their dealings with one another.

Secure electronic transactions

Secure electronic transactions (SET) is a group of standards developed by a consortium including Visa, Mastercard, and several suppliers (see the appendix). It provides a way of authenticating and ensuring the integrity of transactions using credit and debit cards; SET uses the account number, serial number, and expiration date only. Thus, it can be used with any credit card. SET protects the data, but because the card does not have to be present, it does not prevent fraudsters or thieves from using other people's card numbers.

With the advent of smart credit and debit cards, there are moves to enhance SET to include authentication of the card itself, using a smart-card reader attached to the PC. This will provide a much higher level of protection for the retailer and card issuer. Chip cards have recently become available with functions that assist in the implementation of the SET standards.

Specific implementations of SET have already been announced, including Visa's *secure electronic commerce* (SEC). The French banks are moving towards combining SET with their existing *Carte Bancaire* smart cards, and the next stage in this process will be the development of a new French smart card which conforms to the international EMV standards, and which also contains specific SET-related functions.

As described in Chapter 11, other nonbank organizations have also proposed models for securing electronic online payments. Some banks are participating in these models, but typically the banks' view is that the network itself must be made secure before secure payments can be made. Many of the largest North American banks are participating in a combined home banking initiative (using IBM's proprietary Global Network), which they hope can be extended to Internet payments in due course; this initiative does not make use of smart cards.

Other electronic commerce

SET is seen as a platform on which other electronic commerce services can be built. An extension of the standard, known as *secure electronic marketplace for Europe* (SEMPER) is being designed; this integrates SET with the wide variety of electronic data interchange systems used by retailers, transportation companies, and government authorities.

SEMPER raises the possibility of a customer making a purchase online using a smart card; this raises a purchase order from the retailer to a distributor or manufacturer, and at the same time instructs a transportation company to collect and deliver the goods. The customer could also ask for product information or delivery times, which would be available by the same mechanism.

Benefits payment

Fraudulent applications and collection of social security and unemployment benefit are widespread in every country. A system that permits positive identification of claimants offers large savings. Even in countries such as the United Kingdom or the United States, where there is no unique register of citizens, details of claimants (including a biometric) can legitimately be stored and duplicates can be detected.

There have been several pilots of smart-card-based benefits collection, some based on signature verification and others on fingerprints. Fingerprints are preferable in areas where literacy rates are low, and this is a case where security takes priority over customer preferences.

Where a smart card is used for identification, it also offers a payment mechanism: the card itself can be an electronic purse, or the amount loaded may be transferred to a bank account by inserting it into an ATM or telephone terminal. The paying agency saves cash-handling costs, and there is an immediate and complete audit trail.

This model also allows control over the way in which a specific benefit is used. For example, some schools in Britain use smart cards to provide a free school meal allowance to pupils who are entitled to this benefit. A food benefit could be restricted to food stores or a mobility allowance could be restricted to public transportation or petrol costs.

The tools required to implement such a scheme are no different from any other electronic-purse system, while a stronger CVM is available through the biometric.

Loyalty

Many retailers have recognized that retaining and building the loyalty of customers is cheaper than attracting new customers. One of the main tools used to this end is the loyalty card. Customers collect points on their card every time they make a purchase, and this can subsequently be redeemed for goods in the store or from a catalog.

Loyalty schemes allow the retailer to collect additional data about customers' purchasing habits, but they are often expensive both in terms of margin (the value of the points offered) and administration costs. Some schemes operate online (all points are recorded on a central system), while others store the points on a card. Offline points storage on a magnetic-stripe card is subject to fraud, and the storage available on the stripe limits the amount of data that can be collected.

Smart-card loyalty schemes help to overcome these problems: the card can include authentication functions or encrypted data to prevent counterfeit and alteration. Redemption can be controlled by using a PIN. Although points will usually be issued by an online *electronic point of sale* (EPOS) system, the level of points issued and the message displayed to the customer can be varied by using the customer data stored on the card.

Smart cards can also be used as staff cards or discount cards in a single consistent scheme.

Many banks see smart cards as a tool in their strategy to form alliances with retailers, utilities, and other service providers. Here the branding of the card is as much an issue as the services offered. Most of the cards currently being offered have a single function: authentication of the card and identification of the cardholder; points are recorded and redeemed online. An exception to this is the Shell Smart scheme in Britain, although this also avoids security issues by holding only a points count for each subscheme.

Other value-added services

Although it is widely accepted in the banking community, at least in Europe, that bank cards will migrate to chips to avoid the danger of an uncontrollable increase in fraud from magnetic-stripe cards, the cost of making the change makes it difficult to make a business case for the change on this basis alone.

The banks are therefore seeking additional services or benefits that will improve the business case. In some cases these will also help to maintain competition even when the infrastructure is common to all institutions. If the services offer sufficient benefit to the customer, then an extra charge can be made for the card.

Smart cards allow more differentiation than magnetic stripes (the card can store credit limits, customer behavior, language, or cultural preferences), and this in turn can improve customer service. But services such as home banking, the ability to make transactions by cellular telephone (using the GSM standard), and the opportunity to link up with other service providers are seen by many banks as the key to the future.

Developers of bank smart-card systems are concerned that this need for additional benefits may overtake security in the list of priorities. The chips and operating procedures necessary to build a general multi-application, multi-issuer card system with the security demanded by bank systems are not yet fully developed, and it would be unfortunate if the progress made in this direction were jeopardized by demanding too much of the existing systems.

References

[1] *Report to the Council of the European Monetary Institute on Prepaid Cards,* Working Group on EU Payment Systems, May 1994.

[2] *Security of Electronic Money,* Report by the Committee on Payment and Settlement Systems and the Group of Computer Experts of the central banks of the Group of Ten countries, BIS, Basle, 1996.

13

Health

HEALTH CARE AND MEDICAL CARDS are probably the largest application of smart cards in which an older technology is not being replaced. Health care is a highly political subject, and there are wide variations between countries in the extent to which health cards are feasible. Countries also prioritize the applications of the cards in different ways. Nevertheless, the use of smart cards for such a wide range of applications in this sector suggests that there is a common need for the security and storage that only these cards can give.

Insurance

The dominant motive for introducing health cards is control of costs. Although the structure of health insurance varies widely from country to country, each has a common requirement to verify that the person claiming medical services is insured, and to correlate claims for payment with both individual patient records and the doctors' accounts. The majority of cards in the health sector are therefore insurance cards without any medical application.

Health insurance cards do not usually contain a sophisticated cardholder identification method; if they do not contain any medical data, they may well have no cardholder verification at all. The confirmation of identity is expected to be by a separate method, typically personal knowledge of the patient by the doctor or presentation of a national identity card.

The *Versichertenkarte* in Germany is issued to every person covered by health insurance. All of the main insurers have participated, and over 70 million cards have been issued. The main security device on these cards is the general use of write-protected memory; the card can be read in any authorized hospital, clinic, or general practice, but it can only be written to by the insurer. The cards also contain an authentication area which allows each insurer to identify its own cards.

In France, the *Sesame Vitale* scheme has issued over 10 million cards to patients and 100,000 cards to doctors, pharmacists, and other health professionals. Ultimately the system will cover the whole French population covered under the compulsory national insurance scheme. The Sesame cards, which are issued by the *Caisse Nationale d'Assurance Maladie* (National Health Insurance Scheme) serve both to identify the cardholder to the doctor and to provide proof of insurance. Medical records are held separately, although the card is also used to convey prescriptions from physicians to pharmacists. Each card covers a family or *insurance unit*.

Every terminal dials in regularly via the national X.25 network to the social security office to record the services performed or prescriptions dispensed. These records can then be correlated with the claims received from patients. The main aim of the system is to reduce paperwork, but it also contributes significantly to security, reducing the scope for fraudulent claims and ensuring a double-ended check on every transaction.

Medical records

Alternative approaches

Medical records involve large quantities of data. A single patient's records will amount to several hundred kilobytes over a lifetime, and for those with chronic conditions this can extend to megabytes. Storage of this magnitude is probably not economic with current smart-card technology, so several alternative approaches have been followed:

- Storing the data on an online system, using the card to control access to the records;
- Keeping only essential summary information on the card, with references to other records held online;
- Using the card for specific conditions or applications (e.g., hospital treatment or long-term tracking of a condition);
- Restricting the card's application to that of a short-term data carrier, holding only current data and communications between health professionals (e.g., referrals or prescriptions);
- Using a smart card in combination with another card or storage medium to carry the data.

The first of these options (full online storage) is unlikely to be practical in anything other than a very small community. In most countries, patients move freely from one healthcare provider to another, using many different services. One of the main attractions of the card is therefore that it can be held by the patient and used wherever he or she goes for treatment. In Europe, patients can and do move freely from one country to another, and so the long-term requirement is for a system that can be applied throughout the European Union.

The second option (storing only the minimum summary data on the card) is more widely used; this would alert a doctor to special conditions and other treatment that could affect the diagnosis or recommendation. Each of the other approaches has also been used, and each has its own advantages and drawbacks.

Issues

The main issue for a medical data card is the need to protect the *confidentiality* of the data stored on the card and to *restrict access* to qualified health professionals. A common feature of nearly all such schemes, therefore, is the existence of a professional card as well as the patient card. Both cards must usually be inserted into the terminal before the data on the patient card can be accessed. Many schemes require the ability to collect or enter data through a portable terminal (for example, on a home visit or during a hospital round). Special terminals are therefore a common feature of these schemes.

Another problem common to all medical applications of smart cards is the need for *patient consent*. Many people want to understand what is being held on file about them and to control who sees that information. This is not as simple as allowing patients to view the data on their cards; the data will often be in codes and use words not understood by the patient, which may cause unnecessary alarm. A general practitioner will often want to explain to a patient the implications of the data recorded, and this can be time consuming.

With a health card, patients may control who sees the information simply by deciding whether or not to hand over the card. If the card contains administrative as well as health data, then it will be necessary to protect the confidential health data using a PIN or password.

Patient-doctor confidentiality must be carefully maintained. In many cases, the doctor may pass on information to another professional on a need-to-know basis, and generally the patient's consent is not sought or required for this (it is regarded as implicit when the patient agrees to consult the other physician), although most doctors will seek the patient's agreement in the case of particularly sensitive information. Using a smart card or any other electronic system formalizes these relationships and the need for consent, and sometimes removes an element of discretion that can benefit both doctors and patients.

Many of these issues are seen as less important when treating emergencies or chronic illness, and so disease-specific (e.g., cancer care) or emergency (blood group and allergy) cards are easier to implement.

No system can operate without agreement on the *standard codes* to be used for diseases, diagnoses, and drugs. Such a standard does exist, but it is not used with any great uniformity. Doctors, clinics, and hospitals use many different computer systems, often designed for the specific requirements of a national healthcare framework. It will be some years before these are brought into line with the current standards. Large gaps also exist in the chain where some providers have no computer system at all.

Not all medical records can satisfactorily be expressed in terms of codes and dates. It is often necessary to accommodate *additional data,* including text and images. Some of the text data may be quite subjective and yet important to the diagnosis; these records are often sensitive in the context of the relationship between physician and patient.

Another problem with patient cards arises if the patient *loses* the card or it is damaged. It will probably be possible to reconstitute the core personal data on the card (in most cases it will be issued by a central authority or "home" medical practice that has access to the central records), but the other data on the card may have been stored by several practitioners using different systems at different times. The record of use of the card is lost, including, for example, a patient's record of drugs dispensed. Some systems therefore copy all data to a master file whenever the card is used in its home location.

Operational and pilot schemes

Many technologies have been used for health cards. In Japan and China there are large-scale systems using optical cards (see Chapter 4). The large capacity and write-once-read-many-times characteristic of these cards makes them particularly suitable for use in medical-records applications.

The European Union has commissioned two large-scale medical-records card schemes using smart cards:

- *Cardlink* cards will be issued to 250,000 patients in several European countries. The purpose is to test the coding schemes and ease of use as much as to ensure the security and terminal requirements.
- *Diabcards* will be issued to under 1,000 patients with chronic illnesses. This is a demonstration project only at this stage.

In a large-scale French scheme, patients have been issued with *Santal* cards. This is a memory card with several datasets; each dataset can only be accessed by a health care professional using the appropriate professional card for that dataset, and only approved professionals may alter or append data in specific fields. The data is divided into:

- *Hospital data:* Admissions and outpatient consultations;
- *Medical data:* General practitioner or physician consultations, diagnoses, and treatments;

- *Laboratory data:* Blood work;
- *Pharmacy data:* Prescriptions issued and dispensed;
- *Nursing care.*

In the Republic of San Marino (25,000 inhabitants), a smart card issued by the local bank as an ATM and debit card contains a special *health zone,* which can be read by the local hospital, clinics, pharmacists, and medical practices.

A large-scale pilot of a hybrid *opto-smart* card is being carried out in Leverkusen, Germany. This is a single card carrying both an optical memory area and an ISO-standard multifunction smart-card chip. Administrative and short-term data records are held on the chip, as are the keys and secure authentication protocols. Most of the medical data is held on the optical memory, but in an encrypted form [1].

The number of optical readers is minimized in two ways:

- Pharmacists and emergency departments need only read the smart card, as all the data relevant to them are stored there.
- The optical card is read once when the patient arrives at the clinic or hospital, and the data are made available (still encrypted) on the internal network. The patient grants the doctor permission to read the data by entering a password; only the combination of the patient's password and the doctor's professional card allows access to the encrypted data.

The Leverkusen pilot system also features a site card, which is used to initialize the network server and to authorize a professional to use the network, as well as a certification authority (the *trust center*), which authorizes the issue of all of the cards in the system.

Prescription

Several of the schemes already described include a facility for transmitting prescriptions from doctors to pharmacists. There is an increasing need to control access to drugs, both for cost and health reasons, and traditional handwritten paper prescriptions are no longer considered adequate for this task. Printed prescriptions are a slight improvement; however, for full security, an electronic prescription that can be accompanied by a message authentication check is the minimum satisfactory solution. A digital signature could also be added, but this would require a full-scale certification authority arrangement, which has not yet been implemented by any large country.

In the longer term, the certification of health professionals as a part of their qualification process would prevent many cases of fraudulent practitioners and bogus qualifications, and it would permit the implementation of a nationwide or international digital signature scheme.

Almost any smart card, even a small memory card, would permit the operation of a simple prescription-handling system; in addition to preventing prescription fraud, it would form the basis for a control system with a double-ended check on the issue and dispensation of prescriptions.

Patient monitoring

As a further leg in any strategy for controlling costs, the health industry is also seeking ways for patients to monitor themselves to avoid many routine visits. A wide range of machines is available for measuring blood pressure, pulse, sugar levels, and many other indicators in the home. These can often be connected to a home computer or other device; some already have smart-card slots.

Patients on regular drug programs, not all of whom can be relied upon to take their drugs regularly, can be prompted by devices that will dispense the drug, prompt the user to take it, ask for confirmation that it has been taken, and record the whole operation on a card.

Those for whom exercise is a recommended treatment can have their programs loaded onto a smart card; the card is inserted into an aerobic exercise machine (a bicycle, treadmill, or step machine) which also features a pulse monitor. The machine then adjusts the work rate to match the program and records the work carried out.

In each case, the data may be transmitted to the doctor monitoring the program by modem or the patient may take it when visiting the doctor. At this stage, such schemes are nearly always voluntary and there are few security implications. In the longer term, data on these cards should be subjected to similar criteria to those applied to other medical records; although the information is usually less sensitive, the same data protection principles apply.

In general, the data should only be viewed with the consent of the patient by a health professional qualified in the appropriate discipline. The consent may usually be implicit (when the patient hands the card to the doctor or transmits the data to his or her office) or, in the case of more sensitive data, explicit (by entry of a password at the time the data is viewed).

Where such data is transmitted by modem, the link should be established by an authentication routine between the card and the central system (the doctor's surgery or hospital). For maximum security, all data to be transmitted should be encrypted by the card or modem, but this may be considered an unnecessary overhead for this type of data.

Reference

[1] Hartmann, G. et al., "Security Scheme for Hybrid Opto-Smart-Health-Card," presented at *Towards an Electronic Patient Record*, San Diego 1996.

14

Transport

In the modern world, there is so much emphasis on mobility that every system must address the needs of the mobile user. This chapter is concerned with those applications that make mobility itself more efficient. It is the one area where contactless cards have been consistently preferred over other types.

Local public transportation

Organization

In many large cities, all buses, trams, and underground trains are controlled by a single authority, which may be publicly owned. With the growing movement towards privatization, there is a move towards the situation still existing in most smaller towns and rural areas: that of a network of smaller operators, all privately owned and responsible for their own profit and loss. Even within the large metropolitan authorities, many lines or services are run as profit centers and must account for their income and volume of passengers as well as their costs.

Whichever system is used, an efficient public transportation service requires seamless journeys, with passengers traveling on a single ticket across several operators' services. It may also be necessary to have some agreement with a national or local railroad operator if it provides complementary services. All of these systems bring with them the need to apportion revenue between operators fairly.

The system is made more complex in that public transportation is seen as a social service: Some services may be subsidized, either directly or through license conditions. Elderly passengers and children receive concessionary fares. The result is a very complex system, difficult to control. Enforcing fare collection is often uneconomic, which leads to fare evasion. In many areas, both usage and service frequency are dropping.

Magnetic-stripe cards have been used as tickets on all forms of public transportation for many years. Fraud has not been particularly high, but there is concern that forged higher value season tickets, in particular, are now appearing. The machines for reading magnetic-stripe tickets at high speed are expensive to purchase and to maintain. However, the strongest motive for introducing smart cards as public transportation tickets is the extra opportunity they give to collect usage data and to use this to apportion revenue (see Figure 14.1).

There are now many systems operational in which smart cards are used as season passes and for regular users. Some of these form part of *town card* schemes or other multiapplication cards run by a single operator, but the more advanced and complex schemes are usually restricted to the public transportation function.

For the scheme operator, they offer many benefits: Reducing or eliminating the need for cash collection improves security, both for drivers and for the scheme. There is no delay while drivers check or sell tickets. And it is relatively easy to set up fare schemes that reflect the operator's priorities (including, for example, higher fares for night buses or during peak periods).

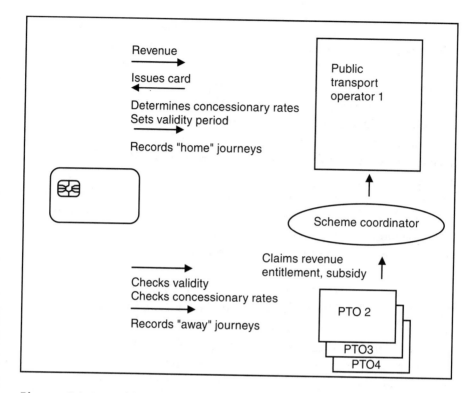

Figure 14.1 Public transport payment and reconciliation.

Types of card

Most public transportation operators prefer a contactless card, primarily for speed of boarding. When people are in a hurry, they become impatient at the time required to insert a card and carry out the transaction; they often remove the card too quickly or damage it physically by mishandling it. With contactless cards, the number of passengers who can board in a given period is much higher. As was mentioned in Chapter 7, contactless readers may also offer higher reliability and ease of cleaning.

Even at the high data transfer rates available with modern contactless cards, the transaction time is still limited by the need to power up the card and establish initial communications. There might also be a need to resolve conflicts between two cards within range of the antenna at the same time. It is found in practice that a 200-ms transaction time is too slow; many passengers have already moved their cards out of range during that time. A realistic maximum seems to be 120–150 ms.

Although reading distances up to 50 cm are possible, most systems use 10 cm or less, and many operators prefer to have passengers bring their card into contact with the reader. This more positive action, confirmed by an audible signal from the reader, results in fewer failed transactions.

The city and province of Seoul, S. Korea, is currently the largest user of contactless card technology. In the city there are 8,700 buses, and in the surrounding province of Kyung Yi, a further 4,300 buses. The cards are sold with an initial value but can be reloaded at ticket sales offices and banks.

The system used in Seoul is also the most common contactless card: Philips/Mikron's MIFARE product. MIFARE uses an application-specific integrated circuit (ASIC) manufactured by Philips. MIFARE includes not only the RF interface and circuitry for handling multiple tags in the field at one time, but also a multiapplication architecture that allows up to 16 different data areas, each protected by two separate keys. The device has facilities allowing mutual authentication between card and reader, encryption of the data being transferred over the RF interface, and an embedded unique serial number per card. The reading system must also incorporate the same proprietary interface, and this is currently the limitation of such systems.

Chip manufacturers are now able to offer a more open solution to this problem (although it is not yet widely used in commercial products): the *combi card*. Combi cards, as described in Chapter 7, have both contact and contactless interfaces, and most of them are based on the standard ISO 7816 architecture and file structure. Security in these cases is provided by the application.

Some operators, including the Paris public transportation operator RATP, have gone for the intermediate technology solution of the *pouch*. The card is loaded in contact mode at a special terminal, but for travel it is inserted into the pouch and used in contactless mode. Maximum security can therefore be applied to the higher value load transaction.

Issues for smart cards

The advantages of control and savings on fare collection are greatly increased if cash and printed tickets can be eliminated altogether, making the whole process automatic. Many public transportation users, however, do not travel regularly on the same routes and indeed may only be visitors to the locality. The system must be able to handle single ticket buyers as well as season tickets and regular users.

Smart card pioneer Innovatron is developing a disposable paper ticket incorporating a chip. The target price for this ticket is under 1 FF (18 cents) in 1998, which can be justified for all but the lowest priced fares. In the longer term, the aim would be to allow open electronic purses, issued by banks or telephone companies, to be used for payment of single fares, but this requires not only the resolution of the "contact vs. contactless" issue but also a wide-ranging set of commercial agreements.

Many public transportation users are mobile: They may visit several different towns or cities. The chances of securing agreement among public transportation operators to honor one another's cards are greater than between public transportation and banks. Such agreements could even extend across borders.

There are two ways to design a scheme involving several operators and a common card: Either the card can hold multiple counters, one for each operator, or each transaction is recorded by the accepting terminal and all reconciliation is carried out on a central system.

The first method involves lower overheads because the card does most of the accounting and less data needs to be transferred to the central system, while the second keeps a more complete and up-to-date record. With the multiple purse system, mobile terminals may be simpler; they may not require any online data transmission at all. Some schemes combine elements of each, as shown on Figure 14.1, where each operator need only keep a record of "away" journeys (those paid for using cards issued by other operators).

The need to keep mobile terminals physically and electrically robust applies particularly to buses, where vibration, dust, and variable electrical supplies are the norm. Again, this argument is to the advantage of contactless cards, as the terminals can be fully enclosed, with no moving parts.

Transaction storage is an issue: Should cards or terminals keep a record of every journey? This greatly increases the storage requirements, but it does enhance the system's ability to recover from faults and to detect fraud and abuse. The usual solution is to keep a small cyclic buffer, of perhaps five or ten journeys, in the card, and a more complete record in the terminal. At the end of a journey, the terminal can be unloaded, either by online transmission or by transferring the data onto a fare collector's or driver's card. The use of driver's cards also has the advantage of completing the cycle of control.

Most of the schemes to date have been either in a single, centrally controlled municipal authority or short-running pilot schemes. Those now being implemented, notably London Transport's Prestige scheme, will have to contend with the longer term issue of the stability of the network of operators.

All of the schemes described here work in "web of trust" encompassing all of the transportation authorities in an area. This is fine as long as the companies are stable and the area is well defined. The main problems arise when companies join or leave the area, when new contracts are established with an additional company in the area, or an operator goes out of business (possibly with debts to the system).

To overcome this, many schemes work on defined time periods: typically one, two, or three months. They attempt to collect all data within, say, five days of the end of the period, and to complete reconciliation within a further three days. In this way exposure is limited and a new company can join at the beginning of the next period.

In the long term, the web of trust will need to be supplemented by a secure multiapplication architecture, where each company is responsible for recording and collecting its own revenue.

Taxis

Modern taxi fleets have radio terminals using private mobile radio frequencies with encrypted transmission of data between the taxi and central system. This is used for dispatching bookings, bidding for bookings, recording positions (either manually or using the satellite-based GPS), and for making payments.

Regular users are frequently issued with account cards; these not only allow the journey to be charged to the account, but may also specify a cost code and identify the specific user on a corporate account. Bank-issued credit cards can also be accepted; the transmission of the card number to the central system is secured using the digital radio system, and the transaction can be authorized online from the central system. Smart-card transactions, whether credit, debit, or purse, can be handled in the same way.

Trains

Railroad systems vary considerably around the world in their level of integration and central control. Since the wave of privatization in the 1990s, some countries' train systems have the same problems as those described for local public transportation. In this case, however, the much wider range of fares, coupled with the need for connections and fares for specific destinations, make the whole process still more complex. Tickets are normally still inspected by a ticket inspector, so it is likely that printed tickets will remain with us for many years yet.

The first areas where smart cards may be seen are for season tickets for commuters; here, the requirement is to allow ticket checking without any delay, and a contactless smart card would provide the answer. Such trains have much in common with metro systems in cities, and the same technology can be used.

At the other end of the scale, smart cards may appear for longer journeys in first class and special trains. Here the requirement is not so much for ticketing purposes as for the inseat entertainment systems currently being designed. These will follow the pattern of inflight entertainment, which is already well established.

Air travel

Requirements

Air travel offers several opportunities and challenges for smart-card vendors. This is a high-profile sector; frequent fliers are a high-earning group, and airlines must expend considerable effort buying and retaining their loyalty. Nearly all have bank cards, and an increasing number will have smart payment cards. International travelers are frequently faced with foreign exchange problems, having to make payments in a currency they do not normally hold. And the whole air-travel experience is seen as a stress that can be minimized by the use of technology.

Against this, the nature of international business means that vendors have to try to meet national requirements in many areas where no international standards exist. The high values involved, the intrinsic vulnerability of the physical controls, and the ease of escape makes airports a magnet for thieves and fraudsters. Systems must be designed in the certain knowledge that they will be attacked.

Electronic ticketing

The first major application of smart cards in air travel seems to be for fast ticketing or ticketless travel.

Airlines have been experimenting with electronic ticketing for some time, but it was only at the end of 1996 that the International Air Transport Association (IATA) agreed a set of standards for ticketless travel using either magnetic-stripe or smart cards. This will not only allow ticketless journeys involving more than one carrier, but will also provide the incentive for major airports to equip themselves with electronic checkin desks.

American Express is piloting a smart card with American Airlines; the card can be used to book a flight or check in by telephone (the booking and seat allocation number is held in the card). At the check-in gate, the details on the card are matched with those on the airline computer. The airline systems will in due course also be linked to American Express' hotel and car rental booking systems, all services being charged to the American Express account. This system uses the IBM MultiFunction Card.

Another priority for airlines is loyalty: Almost every airline has a frequent-flier scheme, under which regular travelers collect points that can be redeemed for future journeys. The German airline Lufthansa is issuing MIFARE contactless cards to its frequent travelers; these allow ticketless checkin and activate the mileage scheme. The card itself is used as an identifier only, with both the electronic ticket and the loyalty bonus held on the airline's central system.

Inflight entertainment

Many airlines now offer a wide range of entertainment and other services at the seat on long-haul flights. These may include videos, flight information, telephony, purchases, electronic games, and even gambling. Many of these services require payment, and the preferred form of payment from the airlines' point of view is a bank-issued credit or debit card. In some cases, the airline has set up a link between its frequent-flier card and a specific bank card; in these cases the airline would prefer that the customer used its own card—this might even be the only card allowed for some functions.

Current handsets will only accept magnetic-stripe cards, but the latest designs are already taking into account the need for smart-card reading. PIN entry, if needed, would require the handsets to be designed and manufactured to a higher standard to maintain the confidentiality of the exchange between keypad and card. Some system designers are considering moving from the present handsets, which have very large numbers of keys, to a simpler cursor control device (more similar to a games controller), which may be easier and more instinctive to use, or even to a touchscreen. Either of these options would require a reconsideration of the PIN entry rules.

Systems must be capable of taking into account local contract laws and rules governing gambling and other services offered. An aircraft is within the jurisdiction of a country it is flying over, and so the rules may change during a flight.

The computer systems required to support such a system are potentially quite complex and costly. The designer must balance the needs of performance and reliability against cost, weight, and the likelihood that the system will be upgraded during its lifetime. As a result, systems are usually highly distributed (see Figure 14.2), and support for card payments (including smart cards) is handled in a separately defined subsystem.

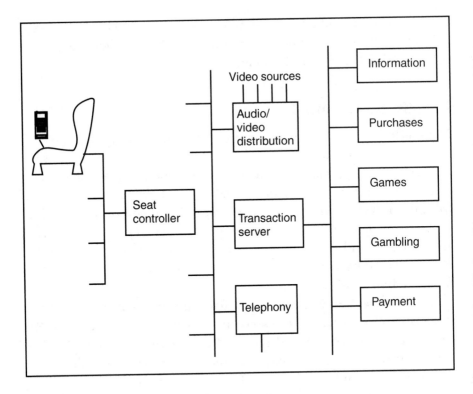

Figure 14.2 Inflight entertainment system.

Road tolling

Traffic growth is an acknowledged problem in many countries, particularly in large cities. With rising incomes and a higher demand for mobility, there seems to be no way to control the growth other than road tolls.

The technology that works best for road tolling is *radio frequency identification* (RFID). Like a contactless smart card, RFID sensors pick up radio signals from fixed antennas and use these signals both to power and to communicate with the fixed system. The systems vary considerably in power and the amount of data they can store or transmit, but those most attractive for road tolling allow a vehicle to pass at normal speed over an antenna embedded in the road.

Most such systems would require the sensor to be fixed permanently to the vehicle. Until all vehicles are fitted with such a unit, this involves

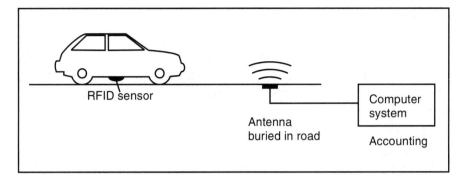

Figure 14.3 Road tolling using RFID.

a significant extra cost. A way around the problem, at least in the shorter term, is to use a contactless smart card, which can be passed by the driver past an antenna alongside the car. This method requires the driver to slow down to 4 m.p.h. or less (there is usually a barrier, but an alternative is to photograph the license plate of cars that pass without paying), but it requires no additional equipment in the car and can be implemented alongside cash-based systems.

Parking

Also related to traffic growth is the difficulty of controlling parking. Many local authorities have problems not only controlling illegal parking, but also collecting the fees due for legal parking. Parking meters and *pay-and-display* stations that collect cash are frequently vandalized. Local authorities want to be able to apply priorities to parking, giving preference to park-and-ride users and encouraging people to avoid peak times.

Such facilities are most easily provided by pay-and-display stations covering a parking lot or street and operated by cards. In some areas, smart cards are being used to avoid the risk of counterfeit or alteration of magnetic-stripe cards. Smart cards allow different groups of customers (staff or local traders, for example) to buy parking at different rates from the general public. Such facilities may only be available during certain hours or subject to restrictions on the length of time or frequency of visits. They may be implemented as part of a more general town-card scheme, offering benefits to residents.

An alternative is to use a smart card in conjunction with a small device incorporating a real-time clock—the personal parking meter. This proprietary device can be placed within sight inside the vehicle, so security is less of an issue. Less "street furniture" is required, and there are no overheads for the installation or maintenance of meters or pay stations. But a very high temperature specification is required for such a device! Since users must buy their own device, this is most suitable for areas where parking is both restricted and paid for.

As with the public transportation system, decisions must be made about the quantity of data to store and collect. Most pay stations operate offline, with a periodic connection to the central system; parking meters are fully offline, unless they use a low data-rate connection along the power system. In this case, it is rare to store any transaction details on the card itself, and even the terminal is more likely to store summaries than the details of each transaction.

There are few security implications for a card used only for parking—it is very similar to a telephone card. Where an open electronic purse or multiapplication card is used, the problems lie more with the card than with the application. These issues are dealt with in Chapter 16.

Where bank-issued cards are used to pay for parking, operators need to ensure that parking-lot and similar low-value transactions cannot be outlets for fraud. Fraudsters sometime use low-value unattended transactions to test whether cards have been reported stolen. Such terminals must have access to a comprehensive hotlist and be able to capture or block the card in the event of a hit.

15

User Identification

Requirements

The underlying function of most cards is to identify the cardholder to a computer system. In this chapter we consider those applications in which this is the only or principal function.

Although the function may be unique, an identity card can be used in many different ways. In an online system, the application itself can be performed by the central system, so the card may often appear to be a multiapplication card. For example, a company identity card could be used for access control, controlling and recording access to photocopiers and fax machines, paying for meals in the canteen, and using the company library.

For open systems, there must usually be two levels of identification: The card must be able to authenticate the issuing organization as well as the cardholder. This implies at least one register of unique organization codes and one common key system to perform the identification.

Thus, such systems can only work with open standards and a centrally controlled register of issuers. Such a register and central control body exists today for only a small number of sectors (banking, telephony, and health).

Other organizations designing identity-card systems usually depend on some proprietary features or arbitrarily chosen codes to ensure uniqueness. Designers would be advised, however, to avoid the situation that now exists for magnetic-stripe access-control cards, where many manufacturers have simply issued card numbers sequentially starting at 00001!

Issues

Level of security

A key decision for an identity card system is whether the card must:

- Confirm the identity of the cardholder before disclosing any data;
- Offer data for confirmation by an external device, system, or person;
- Provide data to the system without any check as to who is using the card.

These correspond to different *levels of security,* and the answer will depend both on the type of threat envisaged and on the extent to which the card must be used offline. The method used for *confirming the cardholder identity* will depend on whether it is being used in a face-to-face or unattended environment.

We must first decide on the level of threat: Are we seeking to block terrorists or casual thieves? What level of false acceptance could be tolerated? Does the card form part of a system or is it the only form of defense?

Then, as described in Chapter 3, we should consider the possible outcomes: What would be the effect if a card were lost or stolen? Would a person finding or stealing a card know where and how it could be used? It is not a good idea to print the address of the building, or even the name of the company, on an access-control card. Would there be any benefit to a thief as a result of being able to use the card?

Online and offline systems

The extent of the system, both in terms of geography and applications, will determine whether or not it is realistic to have all checking carried out *online*. For a school or club building, for example, it is probably possible to have every door connected to a central system. For a national student card system, an online system could not be considered. Even where it is possible, online checking may be an expensive option; the ability of smart cards to operate offline will often offset their extra cost compared with other systems.

There is often scope for compromise: an offline system with exception files (such as hotlists) and key files (for new issuers) updated on a regular basis by modem or by broadcast.

Card issuer responsibilities

The *role of the card issuer* is important in any identity card system, particularly for open or semi-open systems. In any system offered to the public, we have to look at the responsibility and liability the issuer may assume. Could it be held liable if information is disclosed or if a person gains wrongful access to a building? Is the card issuer able to control all aspects of the system or are some reading terminals controlled by others?

Data storage

The level of offline use will govern how much data need be stored on the card. Although smart cards can store several kilobytes of data, they are no match for the hundreds of gigabytes that can be stored on a central system. In most cases, the data on the card must be duplicated elsewhere. Holding the data on the card is, however, valuable if it must be used by many different systems.

For many ID cards, a memory card will be adequate; the card can still be authenticated using a static data-authentication technique, and the cardholder data, including any biometric template, can be stored in data files inside the card. Only where there is a serious threat of counterfeit cards, or where the card must also perform other functions, does it need to be an active device.

Access control

Alternative technologies

Almost every factory, school, and office building seems to need some form of access control nowadays. Smart cards face competition from other types of systems in this application, including biometrics using a central database, barcode cards, and magnetic-stripe cards. This is a market in which the vendors of secure magnetic-stripe cards (such as watermark cards) have had some success.

Smart cards are more expensive than most other forms of cards, but their advantage lies in their ability to operate offline and to perform other functions as well as access control. Many access-control schemes require only very small quantities of cards; this makes printing and personalization very expensive.

In buildings where all locks can be linked to a single computer system, online biometrics are a very attractive solution. This mode of operation does, however, require the system to identify a person from the whole database; this is a slower operation than verifying the identity of one person, and the parameters for false rejection or false acceptance may have to be set more coarsely.

Features

A system must be able to handle visitors; either they must be registered into the central system or they can be issued with temporary cards.

Many building access systems combine different forms of checks. During normal hours access may be by card only, a card with a fixed code may be used at night and on weekends, and the same card with a PIN or biometric may be used for a restricted area. All controls are more effective if they are carried out in front of an operator or security person; in these cases an online system or the card itself can also display a photograph, the person's name, or other characteristics for the security person.

Many cards also have visual identification by means of a photograph or signature on the card; this is not always a good idea, as a person finding the card then knows whom to impersonate.

Other features required by many security systems include:

- Checks that one person cannot hold the door open for another (systems may count the number of people passing and record if more pass than were recorded);
- Checks against a card being passed back to another person waiting to enter;
- Complete records of entries and exits;
- Control of access by groups of cardholders according to the time, presence of a supervisor, or day of the week (including holidays). This type of control nearly always requires an online system.

Special cases

Card-based keys for hotel door locks may operate online or offline. Online locks can be reprogrammed with a new key every time and reset when guest checks out. Offline locks can be reset by an operator, for example when the room is made up at the end of a stay. Simple systems can be operated by magnetic-stripe cards, but more complex schemes including staff cards and access to other hotel facilities are more likely to demand smart cards. The lock can record which key was used to enter and when.

At the Atlanta Olympic Games in 1996, athletes, team members, and staffs were issued with contactless smart cards containing their identification number, height, weight, country, access rights, and a hand geometry template. At the entrance to all restricted areas, the hand geometry was read at the same time as the details from the card; if the card was authenticated satisfactorily, the biometric check passed, and the access rights were valid, then entrance was allowed.

Club membership cards often allow access as well as performing other functions. Where access control is the main function, a memory card with authentication functions is the most suitable type of card. If other functions such as cashless payment are to be included, the memory card may still be adequate if the system can operate fully online. Otherwise, a microprocessor card should be used.

In most other situations in which they are used, smart cards also perform other functions. Access to a computer system or cashless payment are the most common.

Other applications

Universities and schools have been popular with smart-card vendors as test sites; it turns out that the technology offers some real benefits in this sector. In most cases, the card functions only as an ID card; the use of smart cards makes the cards themselves more secure through the use of authentication.

Increasingly, however, the ID card is being combined with a cashless payment function. The card can be used in canteens, bars, and vending machines, and it can be used to pay for photocopying and other services on campus. It can also record attendance at lectures offsite and keep track of course credits; storing these on the card allows the student to check his or her records using a kiosk or offline PC system.

In many British schools, pupils use their smart cards to record their attendance, to pay for school meals and vending machine snacks, and to gain access to computer rooms and other restricted facilities. The card allows pupils eligible for school meals allowance to receive it through the card, without other pupils being aware who is receiving free meals. In some instances healthy eating is also promoted by awarding points according to the type of food being bought; these are stored on the card, and the school can decide its own reward system.

Smart cards could also be used as passports or national identity cards; realistically they are more likely to be used to supplement a printed card for regular travelers or other specific functions. The European Union has asked member states to make provision for a chip on any new drivers' license cards; the obstacles to introducing such cards are more political than technical, although several standardization issues would also have to be resolved.

Many of the more futuristic scenarios dreamed up by technical journalists and technology enthusiasts involve personal preference cards: cards that record the way I like my coffee, what type of background music

I prefer, how I set the driving position in my car, and so on. Here the obstacles are not technical or political but organizational: Who would issue such a card and set the standards for recording preferences? It is likely that such a scheme will take off first within one specific sector, where a suitable coordinating body already exists and will spread from there by commercial alliances and cobranding.

Such a card must have the ability to record data under many different headings, probably in a very wide but shallow file structure. Because cardholders have different preferences for confidentiality and convenience, access to each field must be controlled by both the application and the cardholder. This could be handled within the existing ISO 7816 file structure, but it would not be efficient; a further generation of file structures may be required for this dream to be realized.

16

Multiapplication Cards

THROUGHOUT THE HISTORY of the smart card, writers have assumed that a single card will be able to perform many functions, across the whole range of applications discussed in the last few chapters. In practice, few such cards exist; this chapter explores the obstacles and the means by which they may be overcome.

Functions and applications

Many cards can support what seem to be multiple applications (such as access control, vending, library lending, and photocopier control) without needing to have more than one application (program) loaded on the card.

In the case of a microprocessor card, the card would have one *application* loaded, but this application would include several *commands:* for example, authenticate card, authenticate terminal, send ID, increase purse value, decrease purse value. The application would, however, be

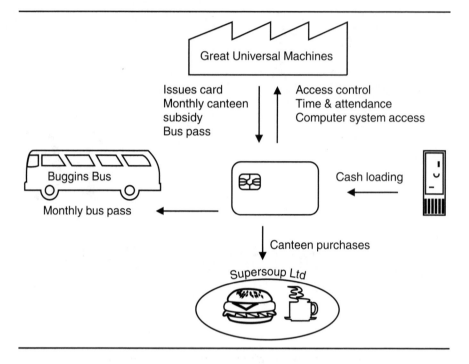

Figure 16.1 Great Universal company card.

developed and tested as a single unit, often as an integral part of the card's mask.

A memory card could be made to perform the same functions, but in this case the commands are executed by the terminal rather than by the card; this implies a higher level of control of the terminals where the cards will be used.

For an example, our company, Great Universal Machines, uses Supersoup Limited to operate the canteen, and bus services to and from the factory are provided by Buggins Bus Company. Great Universal decides to issue smart cards to all employees; these will be used for registering time and attendance, for access control out of hours and to controlled areas, and for logging on to the computer network. To give better control of the subsidized canteen and bus services, each employee receives a canteen subsidy each month, and money can also be loaded onto the card by a voluntary deduction from the salary or at a card loading

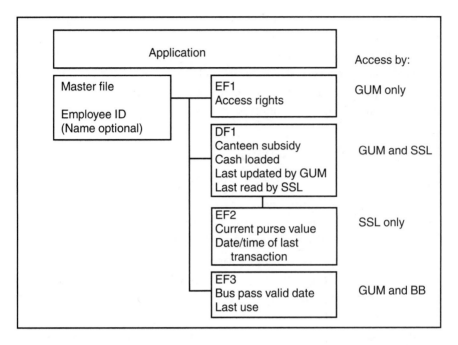

Figure 16.2 Card structure for Great Universal company card.

station. The bus pass is also revalidated at the beginning of each month (see Figure 16.1).

All of this can be handled by a single application on the card, yet still give adequate control to Supersoup and Buggins Bus (see Figure 16.2).

A group of South African banks has developed a single application for payment cards that would support debit, credit, and purse functions. Even using the EMV standard for payment cards, this same range of functions could be supported.

So when is a multiapplication card really needed? When several companies are responsible for the different applications, when the applications must be developed independently, and particularly when we may need to download new applications onto the card after it is issued. In this situation we need to protect each application and its data from the others.

We could also envisage situations within a single organization where the implementation of a smart-card program is phased; funding may not be available for all applications from the start, or some applications may simply have a longer development timescale. In these cases, the requirement for independence is more technical than commercial.

We must also look to the generality of our solution and to the standards used. Many systems achieve at least some of their security by relying on proprietary features. For a truly multiapplication card, however, open standards must be used.

Card operating system

Most of today's smart cards do not have a true operating system, in the sense that larger computer systems use the term. There are no external resources to control other than the I/O port, and only one application runs at a time. In the longer term, there is no reason why this must remain true; with faster processes and more memory, applications could be suspended or run in contention for the I/O port, cryptoprocessor, or other functions contained within the card. This would require a more conventional operating system with all of its overheads.

With current card masks, programs are not held in the same memory as data; interprocess protection is thereby made much simpler than with, for example, a PC system. It is easier to ensure that one application does not affect the integrity of another application; the operating system must simply ensure that only one application can gain access to its data, with the exception of data that is explicitly shared.

This is where the problems begin. Defining a mechanism that allows one application to share its data with another authorized application, probably in specific ways (e.g., application A creates the data field, B may write and read the data, C may only read it), requires an operating system designed for that purpose.

Where, in addition, the data fields are protected by keys, the key management of the card must be extended so that applications can manage their own keys. (In most smart cards, there is one key file, which is used to give access to all data fields). It may even be necessary for two applications (but not necessarily all applications on the card) to have access to the same key. This requires a comprehensive scheme for granting access to an elementary file (EF)—or possibly even to fields within that EF—according to the application and access control conditions to be met.

Such a scheme involves considerable overheads both on data storage and on processing power, and so it can only be implemented on microprocessors at the top of today's memory and power range. Masks of this type are starting to appear. To be effective, they must be provably secure (using ITSEC or similar criteria—see Chapter 3), so that one application issuer can have confidence that the card will protect its application completely.

Downloading

Cards that can run multiple applications offer the opportunity to load further applications onto the card after it is issued. This would allow applications to be updated or additional services to be offered to cardholders.

It would even allow the scenario in which I purchase a card (in much the same way as I would a PC) and then ask my preferred service providers (my bank, one or more transport companies, a telephone company, or my employer) to load *their* applications onto *my* card.

In practice, cards today and for the next few years do not have enough memory or internal protection to allow any generalized downloading capability. However, downloading could be used within a restricted circle of application issuers in the context of a cobranding or cross-selling agreement. Examples would be public transportation companies in adjacent areas, or a bank, insurance company, and stockbroker.

In this environment, the card must authenticate both the application issuer and the downloading station before it proceeds to download any application or data. A strong authentication scheme should be used, and it may well be necessary to carry out the whole operation in a secure environment or online to a trusted system. To authenticate the application issuer offline, the card must carry the public key of a certification authority (which might be its own issuer) so that it can verify the certificate offered by the application issuer. The terminal should also be authenticated to avoid any danger of a subsequent replay.

Soon, there will probably be viruses for downloadable smart cards. The risk of such viruses causing significant damage can be minimized by the use of a secure operating system that protects applications from each other. A card that has just had a new application downloaded should not run that application until it has had a chance to run a virus check. This would involve the card connecting online to its issuer or to another trusted system for a check of the loaded application.

Hybrid card types

Several multiapplication combinations involve a public transportation or access-control application combined with a financial application. As we have seen, issuers in the first group have a strong preference for contactless cards, while those in the second are committed to using contact cards.

Where *combi cards* are used to overcome this problem, we must also consider whether it is appropriate for the applications to share memory—and, if so, whether the access control to each section of memory should be the same. Unless there is a need for the two applications to communicate, the answer is probably no: they should be kept separate. An example of such a split-function card is Mikron's MIFARE Plus, which combines an ISO7816 microprocessor card with its standard MIFARE product.

If this is not done, the security of the card, and hence of the system, suffers from a least-common-denominator effect: Each aspect of security is only as strong as the weaker of the two systems.

Card control

Where applications are updated, it is important to track (within the issuer's card-management system) which applications and versions are held on each card; as with all software, version mismatches can cause problems. Keys are often updated along with applications, and if an old set of keys is in use, this may compromise the security of other, newer applications on the card.

The most common approach today, when the use of multiple applications is limited and version updates generally infrequent, is to load all of the relevant applications onto every card and to enable and disable them when required. This allows all applications to be kept in step and only requires one set of flags in the card issuer system.

Issuer responsibilities

One factor that will deter many card issuers from opening up their cards to other applications is the fiscal and contractual responsibility that the card issuer may take on. This will partly depend on the law in each country; for example:

- In Denmark, there is a specific law for payment cards but not for other cards.
- In Germany, the card issuer is deemed to have a contract with the cardholder by virtue of issuing the card, but this would not apply to other application issuers.
- In the United Kingdom, the card issuer's responsibility includes liability for the goods purchased in the case of credit cards, but not for debit or purse cards.

In general, the card issuer will have some duty to ensure that the cardholder is not exposed to financial loss or exposure of confidential data as a result of normal use of the card. It will also have a responsibility towards those accepting the card (in payment or as proof of identity) and towards other application issuers.

If, for example, a card carries the name and account number of the cardholder, does this imply that any form of credit check has been carried out or is any guarantee of payment made or implied? Card acceptors in the retail and service industries are used to the terms of the financial card schemes, which do guarantee payment for validly performed transactions.

Issuers of multiapplication cards will need to look carefully at the legal liability they could be accepting and ensure that the cards and mechanisms they propose will provide adequate protection.

Consumer issues

Cardholders who receive a card by virtue of their employment, as benefit claimants, or as pupils at a school are entitled to protection of their personal data in the same way as other citizens, but are otherwise unlikely to have any say in how the card is used. Issuers of such cards will generally restrict the data stored on the card and also limit the card to applications under their direct control.

Where the cardholder has chosen to have a card (whether it is issued free, as part of a service, or directly purchased), he or she may want to know what is stored on the card and how it may be used. In the case of a multifunction card, it may be difficult to answer either of these questions, particularly if no one organization has full control of all of the data. Usually the card issuer retains ownership of the card itself; for a multi-function card this may no longer be appropriate.

There is a special case in which a government decides to issue a national identity card, drivers' license, or other mandatory card. Multi-functionality in this case is almost certainly taboo in most western countries (although it would be the ideal medium for a totalitarian state). It is essential to define the data and its use before the card is issued.

Interchange and compatibility with existing card systems

A further set of problems arises when a multiapplication card is introduced to an environment that had been single-application. Assumptions made about keys and certificates will often no longer be valid, and both card and terminal must carry the additional data needed to identify application issuers as well as card issuers.

Although the latest generation of card operating systems takes us much closer to a full multiapplication environment, the migration of applications will take many years.

17

Current Trends and Issues

Market forecasts

Europe currently accounts for around 80% of the smart-card market, with perhaps 16% in Asia and 3% in North America. However, the market outside Europe is growing. Several Asian countries have large smart-card programs; Singapore, Hong Kong, Taiwan, and China all have major projects. Even in North America, where interest in smart cards has been very slow to take off, around five million cards were issued in 1996.

Estimates of the potential growth of the smart-card market vary widely. At one end they are simply projections of existing markets, growing by 15–20 percent a year. Those at the other extreme are pure fantasy. A middle view, after eliminating the extremes, would suggest that the market could grow from around 500 million cards a year in 1996 to 2.5 billion in 2000. Telephone cards will still remain by far the largest application by volume, accounting for around half the total, but the higher

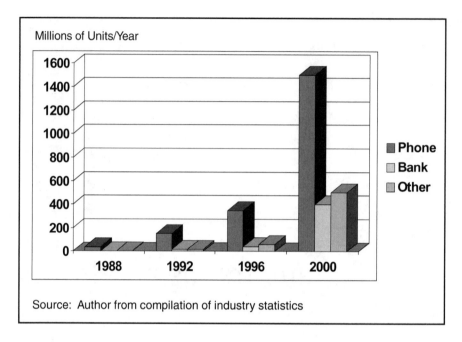

Figure 17.1 Forecast smart-card market growth.

value bank and multifunction cards, with up to 400 million cards, will account for more than half of the market value.

In the short term, the market for closed systems (in companies, schools, and universities) is likely to grow faster than that for full-scale public systems. Closed systems are easier to manage, and it is easier to pinpoint and control potential points of weakness. However, many designers also incorporate proprietary features that appear to improve security. As we have seen, this is not always a safe option, as the level of care and expenditure that goes into the design of the proprietary feature is likely to be less than for the corresponding commercial product.

In the longer term, the emphasis is on multifunctionality. As we have seen in Chapter 16, multifunctionality does not necessarily mean multiple applications on the card or multiple application issuers. However, if the commercial and contractual issues can be resolved, there will be a demand for cards that can support multiple issuers on one card, simulta-neous application running (e.g., running a biometric identification or a financial transfer application from within another application), and downloading, all the time protecting each application and its data from other applications and from the outside world.

Not all users will want powerful multiapplication cards: Many people see multiple cards as being easier to manage than single cards. The pressure for multiple applications comes from card issuers and system designers; the authentication and identification features occupy most of the space on the card, and it is then easy to add additional functions at very little marginal cost. The cost of the card can then be spread across a wider range of applications.

Cards

Chips

A minimum specification for cards is establishing itself, driven by the large-volume telephone applications. Future cards are likely to have a minimum of 2 kb (256 bytes) of memory, with a basic level of memory protection and PROM-based card identification, which can be used for authentication.

High-end designs for smart-card chips are pushing upwards: 1μ technology will shortly give way to 0.5μ, and then to 0.35μ, and 0.2μ. The driver for this is volume: semiconductor manufacturers have lines for each minimum feature size but are unlikely to allocate the highest specification line to a low-volume product. Voltages are moving down; few chips now require the 21V E^2PROM programming voltage, and the main supply voltage is now commonly 3V rather than 5V. The next step is 1.2V. Memory densities and processor speeds are rising each year, broadly following Moore's Law.[1]

Researchers are now considering the limits to this process: as the semiconducting layer becomes thinner and the individual features narrower, we start to reach the level in which the number of free electrons becomes a significant factor. Current and voltage can no longer be regarded as continuous but rather the movement of discrete units of charge.

---

1. Gordon Moore, one of the founders of Intel, observed that the memory size available on a single chip doubled every 18 months.

Conventional wisdom still suggests that the maximum size for a smart card chip is 20–25 mm^2. Some manufacturers are breaching these limits, particularly for contactless cards, and it seems that for certain applications cards with these larger chips are reliable.

Most current smart-card microprocessors are 8-bit designs. Sixteen-bit designs are not common; designs seem to be moving directly to 32 bits, and these processors are now starting to appear. Several manufacturers are also looking at reduced instruction set (RISC) architectures for specific applications requiring higher performance.

Manufacturers are currently predicting that by the year 2000, 0.25μ processes and 1.2V operation will be the norm; memory sizes for flash memory cards could be several megabytes, although these will only be suitable for some applications. On perhaps a slightly longer timescale, FRAM could supplant E^{2}PROM completely in smart-card chips; this would allow an order of magnitude increase in memory capacity for standard microprocessor chips—up to 256 kb or even 1 Mb. Where E^{2}PROM does continue to be used, it is becoming much faster; write cycle times of 1 ms will be achievable by the end of the century.

From a security point of view, chip designs become ever more complex, and more of the tricks hitherto applied only to top-of-the-range chips are now being used in standard designs. More chips incorporate detection for out-of-specification operation: temperature and rate-of-change detection are being added to existing checks on voltage and clock speed. Some chips now generate their own clock, synchronizing it from time to time with the external clock. This removes the opportunity for hackers to analyze chips by operating them very slowly or to cause errors by introducing clock "glitches." Out-of-specification detectors are becoming more reliable, so they are less likely to be ignored by software designers.

An increasing proportion of microprocessor chips carries a cryptographic coprocessor, and this allows mask designers to incorporate public key cryptography as a standard feature at acceptable speed. The current state of the art allows 1,024-bit RSA encryption in under 500 ms; this is still slightly too slow for many mainstream applications such as electronic purses. However, it is acceptable for applications such as access to safes or computer systems, where security is the main function.

An example of what can be achieved by current technology is given by the *cryptographic reduced instruction set processor* (CRISP), developed by the SOScard Consortium—General Information Systems in the United Kingdom, DigiCash in the Netherlands, and Graz University of Technology in Austria. This development was partially funded by the European Commission under the Open Microprocessor Systems Initiative. It is a RISC design with an integrated cryptoprocessor. The data processing speed is 20 million (8-bit) instructions per second (MIPS), and the cryptographic engine is optimized for public key encryption, with a 640-bit RSA signature in 50 ms.

The card's security features include an internal clock, dynamic allocation of internal functions, and privileged instructions. The card also incorporates a full range of memory protection and session-management functions to ensure the integrity of multiple applications running in a true multitasking mode. Although this was a research project, chips based on this design are likely to be available for commercial applications in 1998.

Masks

As crypto engines become more efficient, mask designers have also become more skilled at making use of these features to implement fast public key encryption; the most common use for this is for two-way authentication of card and terminal at the beginning of the transaction. High-end chips are now able to generate their own RSA key pairs on the chip, so that the secret key never leaves the chip.

Many masks now incorporate interprocess protection as a standard. This enables applications to be developed independently, often using external development kits. The development kits themselves are now being seen as an important part of the product: A good development kit enables systems house personnel who are not specialists in smart cards to make use of these products.

The more advanced operating systems now offer more complex access-control mechanisms than traditional cards. As with conventional operating systems, applications can be given the right to read, write, create, update, or delete data within elementary files, while the right to run an application may require authentication of the terminal, a connected process, the cardholder, or any combination of these.

Some masks also include specific features to allow recovery from incomplete transactions. As we discussed in Chapter 16, recovery in a multiapplication environment can become very complex, and it is an advantage if this can be handled in a standard way by the operating system rather than requiring each application to build its own transaction log files for use in recovery.

More mask features are now being offered as standard by the semiconductor manufacturers. This enables them to optimize the protection for their own hardware, and indeed some of the hardware and software must be designed in concert. Where a card is advertised as having certain security features, it may be necessary for the manufacturer to produce software that takes advantage of these features rather than leaving the implementation in the hands of the purchaser.

Contact/contactless

There is still an issue within the industry over contactless cards. While proponents of contactless cards see them as the only way to allow the fast transaction times required in public transportation, access control, and similar applications, other sectors still have doubts over the security of the radio interface and over the limited number of sources of supply. Combi cards are more expensive and in general the security offered is still lower than that of the better contact cards.

Application downloading

It is still unclear how or whether the market will develop towards the truly "open" card, with applications freely downloadable. One set of problems still to be resolved is the ownership of the card and the liability of the card issuer for any problems with subsequent applications.

In general, for any application-downloading scheme to work, there must be a mechanism for authenticating application and terminals. This in turn will require the creation of registers for authorized issuers, applications, schemes, and terminal operators. This requires a centrally designated authority in each country or sector within a national framework. In Britain, the Department of Trade and Industry has taken the initiative in this area and proposes to set up a central register of issuers, applications, and masks.

To overcome the problem of downloading control, several companies have produced interpreter-based cards. These allow applications to be downloaded and to run within a controlled environment. This concept may be extended to include short-lived *applets* using languages such as Java; cards capable of running Java have been designed but none is used in a commercial scheme.

Downloading opens up the possibility of the security of a smart card being breached by the use of a *Trojan horse*. Systems available for detecting the presence of a Trojan Horse or virus on a smart card have not yet been developed.

Encryption

Smart-card systems make use of a limited range of encryption products, usually from commercial suppliers or specialist encryption developers. The strengths and limitations of these products are usually well understood by their developers. System designers often do not share this knowledge and are unsure how to make use of the products or to avoid compromising their security.

Within several sectors, notably telecommunication and banking, guidelines are now available for the use of encryption. Some of these take the form of international standards, others are published industry standards or de facto standards set by a dominant supplier. This should make it easier for systems house personnel and other nonspecialists to make use of the products.

International issues

In several industries, the existence of national standards makes the adoption of international standards more difficult. Yet several of these industries, including banking, are those in which there is a pressing need for international cooperation. But some of the most advanced smart-card schemes in the banking industry are national implementations, and there is a strong argument that waiting for international standards simply gives a bank a competitive disadvantage in the national market.

The image of the smart card may therefore suffer when users find that they are unable to use their South African cards in British, Dutch, or French smart-card terminals, or that the facilities available to them are much more limited. The appearance of the EMV standards has only gone a short way towards correcting this problem.

Even in those areas where the European Union has taken a strong position (such as health cards), national schemes remain the norm. The legal and administrative framework in each country is sufficiently different for the applications and mode of operation to be quite distinct.

Single terminals

In sectors in which there will be many card issuers or schemes (as in banking, car parking, or airline ticketing), card acceptors will insist on one single terminal accepting all cards. The ISO 7816 standards do not guarantee this level of interoperability, and there is a clear need for each sector to define common standards, taking into account the need for future upgrades and additional functions.

As with the cards, several companies have proposed interpreter-based terminals (such as Europay's Open Terminal Architecture for financial terminals). However, others have felt that this architecture is too restrictive, and the general approach has been to divide the implementation into several layers, so that an upgrade would in general only affect one layer.

Where there is a possibility of new scheme operators being included within a framework, cards and terminals must have sufficient capacity to handle additional keys and possibly additional SAMs within the terminal. The use of SAMs makes it difficult to specify generalized systems that can be upgraded in the future.

Standards

An absence of widely accepted standards has dogged the smart-card industry from its infancy. International standards were very slow to arrive. Even now that more standards are accepted, they only cover some characteristics, so that there is in practice very little interoperability between different manufacturers' systems. Every manufacturer uses a proprietary development environment (some even have more than one), and none of these is familiar to programmers brought up on modern high-level languages and open systems concepts.

The effect of these nonstandard environments is to keep smart-card developments out of the mainstream of computing. Although many of the companies involved in the smart-card industry are highly innovative, the number of innovative companies that feel excluded from the market is far higher.

The ISO 7816 family of standards is being enlarged and updated. Future parts of this standard will include the use of a smart card as a database storage medium. Increasing numbers of sectors now have some form of standards covering their applications. Several large telephone companies in Europe, North America, and Asia have created the Global Chipcard Alliance to promote standards and share development experience in the introduction of smart-card technology.

In an effort to promote convergence between chip card and PC standards, a group of five major players (Bull CP8, Hewlett-Packard, Microsoft, Schlumberger, and Siemens Nixdorf) has developed a set of draft specifications known as the PC Smart Card (PC/SC) standards. These are freely available to developers and are intended to facilitate the interoperability of smart cards used in the PC environment.

IBM is promoting the use of its OpenCard framework for the use of smart cards in Network Computers (NCs), and there is talk of attempting to merge this framework with the PC/SC standards.

Because of the high level of interaction between the chip hardware and the security features of the card, it is difficult to establish the level of independence between hardware and software to which mainstream designers have become accustomed. One proposed solution is the use of interpreters: Europay proposed a version of Forth for EMV cards, and Gempus and Schlumberger have announced that they are cooperating on a card that will use the Java language. Applications written in Java are likely to be too large and too slow for the current generation of smart cards, but these manufacturers clearly intend that future generation cards should provide both the memory capacity and the speed to handle this standard.

Market structure

Perhaps the most important movement, however, is just beginning: As smart cards begin to be incorporated in other schemes, alongside "normal" computing standards, mainstream system designers and programmers are becoming aware of the possibilities of the technology and are starting to apply their own, much less rigid, rules to the technology. De facto standards are cheerfully accepted and the criterion for selecting a product is its ease of use rather than its international standardization status.

The smart card market has always been vertically organized, with at least five layers (see Figure 17.2). Until recently, the number of players in the top three layers was closely controlled by the patent holders; many systems integrators were deterred from working in this market by the need to pay license fees and accept controls on their activities. With

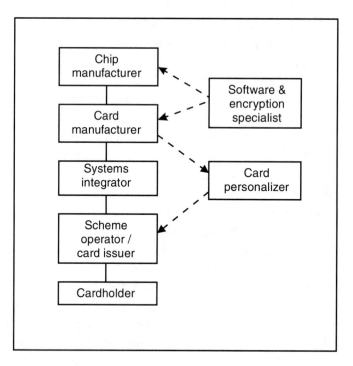

Figure 17.2   Smart card market structure.

the expiration of the main patents, there is scope for a new market structure to arise, with a much wider range of companies performing the system integration role and with greater involvement by end users.

This should allow a broader range of companies with experience of the security requirements of specific sectors to enter the market and to make use of their experience in designing smart-card systems.

18

Security Model

Aims

The reader will by now be aware that there is no such thing as perfect security. A security design is a balance between the requirements of confidentiality, authentication, and integrity on the one hand, and convenience, cost, and reliability on the other (see Figure 18.1). The only way to ensure that no one can ever read your confidential data is to destroy it completely.

The more realistic aim of every security system should be to ensure that *the effort required to break the system exceeds the reward.*

Reducing the reward

We reduce the severity of the risk by:

- *Limiting the part of the system affected:* Systems should be subdivided so that no one key gives access to the whole system,

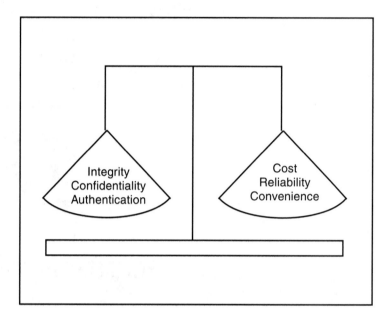

Figure 18.1 Security in the balance.

and that loss of one piece of data only affects a limited part of the system.

- *Limiting the time for which the system is exposed:* Exposure can be greatly reduced by reducing the life of keys and other critical data.

- *Limiting the amount at risk:* The value attributed to one card should be limited according to the strength of the security mechanisms used.

- *Reducing motivation:* Avoid setting the system up as a target by maintaining an atmosphere of trust combined with control and avoiding high-profile claims.

The last element is particularly important in deterring hacking attempts because for many hackers the intellectual challenge, and the publication of their achievements, is its own reward.

Increasing the effort

The effort required to break the system can be increased by all the methods discussed in this book:

- *Focus on the data at risk:* Do not increase the security of unimportant areas unnecessarily (unless as a deliberate decoy); it will distract attention from the more important areas.

- *Set security objectives:* These must be consistent and realistic. Do not demand military level security unless you can afford it.

- *Choice of card type:* Select a card with a suitable level of security; specifications are rising all the time.

- *Encryption of data:* Use commercially available products, but ensure that they were designed for the purpose you are seeking.

- *Card, terminal, and system authentication:* These will be discussed later in this chapter.

- *Cardholder verification:* Use PINs, passwords, or biometrics.

- *Key management:* As soon as you start to use encryption, you also have to manage keys. Use commercially available systems where possible and follow the procedures recommended.

- *Control of access rights:* Access rights must be controlled both within the card and in any associated computer systems.

- *Physical security:* This remains important even where automatic checks are in place.

- *Organization:* Tight organizational control is essential to any secure system.

- *Procedures:* Procedures must be written and tested.

- *Audit and external verification:* The system should be independently reviewed and monitored.

Criteria

It will by now be clear that smart-card system security is not a single-dimensional effect, and that each case must be looked at on its merits.

Nevertheless, as a starting point, we propose in this chapter a sequence of requirement setting, design, and analysis to assist in designing and assessing any system that seeks to use smart cards to enhance data security.

Types of security

The first step is to determine the broad requirements for:

- *Confidentiality:* What data must be kept confidential and from whom? What is the value of the information in the wrong hands?
- *Authentication:* How important is it to ensure that the card being used has been issued within our scheme? Must we also authenticate the person or is it the card that matters? Where the terminal can alter data on the card, must the card also authenticate the terminal?
- *Integrity:* What would be the effect if data were lost or altered during transmission and storage?

Ideally, we should go on to set quantitative criteria for each of these areas as described in Chapter 3. However, we can use a simple model to construct the outline of a system; this model can then be analyzed and the appropriate strength of protection applied to each element.

Model

The simple model assumes that data must be stored, transmitted, and used. At each stage we must apply the appropriate level and type of security mechanism (see Figure 18.2).

Storage

We start with the requirement for data storage. If the data must be portable or accessed offline, we can store it in a smart card; otherwise, it will be preferable to store the data on a computer system and use the card to gain access to that data.

According to the level of confidentiality required, we can:

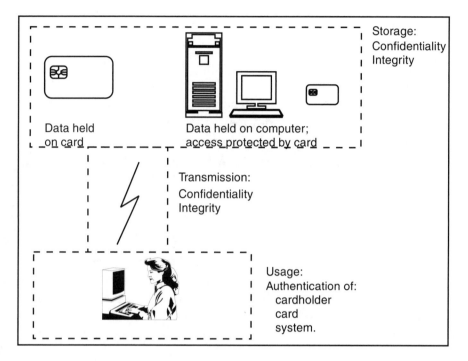

Figure 18.2   Security model.

- Store it in clear on a memory card or a computer file;
- Store it in a protected memory card or on a computer with access control rights controlled by a password or card system;
- Encrypt the data or store it in a secure memory card.

Data is made unintelligible to unauthorized persons through encryption with secret keys, usually with symmetric algorithms. We can increase the strength of the encryption by increasing the length of the key.

We must also check that data stored are not altered, and that data cannot be lost as a result of a malfunction, misoperation, or power failure. Again, there are several levels depending on the degree of integrity required:

- At the lowest level, data in the card are assumed not to change except during an application. A checksum may be held for computer files. Daily backups are made from all central files, and transaction logs are kept for several days.

- For added security, checksums can be associated with all data. Every data item that changes is logged; transaction logs within the card are usually quite small (5–10 transactions) but on a central computer system they may be kept for weeks.

- For demanding applications, MACs are associated with critical fields. Important data are also held in shadow files on the central system.

Transmission

When data are transmitted from one system to another, we must ensure that the information is not altered, accidentally or deliberately, on the way. If it is confidential, then we must protect the interface either physically or by encryption. We check the integrity of messages by using *cyclic redundancy checks* (CRCs), transaction counters, and MACs:

- A message CRC is the most basic check; it must be long enough to pick up any accidental change, but it provides no protection against deliberate alteration.

- A MAC that includes the message number, date, time ,and key data elements provides good protection against alteration.

- To include checks for nondelivery, not only the data but also the acknowledgment should be MACed.

Use

When the data is used, we must perform checks to ensure that the person presenting the card is the cardholder. It may also be necessary to check the authenticity of the card and of any terminal or system in which it is being used.

- For many systems, possession of the card is sufficient to give access or to allow its use. Cards carrying a signature or photograph can be checked manually in face-to-face environments.

- The simplest form of automatic cardholder authentication is by a PIN. Although this method has limitations, its ease of implementation, customer acceptance, and proven record weigh strongly in its favor.
- For higher security applications, several effective biometric checks can now be implemented through smart cards.

When it comes to authenticating the card itself, there are again several options:

- The simplest method is to check the card identification held in PROM. This is normally only suitable for closed systems.
- For open systems, we must store an individual cryptographically derived checksum in the card, or resort to public key cryptography. A simple asymmetric scheme does not require the card to perform public key encryption.
- Where the card can perform public key encryption fast enough, we can use two-way dynamic authentication using a zero-knowledge protocol. This is the most secure way of authenticating a card.

Finally, we must consider the need to authenticate the terminals and systems in which the card will be used.

- Again, there is a basic check in that the cardholder does not offer the card unless he or she believes that the terminal is authentic.
- The card may supplement this by checking a simple symmetric or asymmetric key certificate held in the terminal.
- For maximum security, two-way identification should be used; this is particularly important when operations that change data in the card may be performed offline.

The model is summarized in Table 18.1, where we list a range of typical tools that may be used to meet low, medium, and high requirements for each of these conditions.

Table 18.1 Security Requirements and Tools

Parameter	Level of Requirement		
	Low	**Medium**	**High**
Confidentiality	Data stored in clear on memory card or computer	Data stored in protected memory card or on computer with access control	Data encrypted or stored in secure memory card
Authentication (cardholder)	None (possession of card is entitlement)	PIN	Biometric
Authentication (cards)	Check card identi-fication in PROM	Individual cryptographically derived checksum held in card (SDA)	Two-way dynamic authentication (zero-knowledge protocol)
Authentication (terminals and systems)	None (cardholder does not offer card unless satisfied)	Symmetric or asymmetric key certificate held in terminal	Two-way identification, both for offline and online operation
Integrity (communications)	Message CRC	MAC, to include message number, date, time, and key data	MAC as left, with acknowledgments
Integrity (storage)	Checksums for data held on computer files only; daily backups	Checksums associated with all data; full transaction logging	MACs associated with critical fields; important data also held in shadow files

Analysis

Initial situation analysis

Having ascertained the level of requirement in each area and the outline of the system to be used, the next thing to consider is the pattern of risk. Risk analysis was considered in more detail in Chapter 3. We must consider a very wide range of outcomes, both positive and negative, and assess for each the severity and likelihood of the outcome.

If there is any gain to be obtained by attacking the system, then the designer must assume that attacks will occur. He or she must consider the likelihood and possible severity of each of these attacks: How persistent or well motivated would the attacker be? What resources could they have access to? How much access would the attacker have? Would it depend on a range of coincidences or would the opportunity always be present?

Sources of attack

All possible sources of attack should be considered, for example:

- *Normal cardholders* will try to do things that the system is designed to stop them from doing or they may use the system in ways the designer has not considered. Often such attacks are not malicious or even intentional, but they are seen as attacks by the system.

- *Careless cardholders* often lose or damage cards or terminals. Often some of the card issuer's best or most important customers are careless, and customer-service requirements dictate that the system must be able to recover quickly from this situation to restore the cardholder's rights.

- *Malicious or untrustworthy cardholders* have the highest level of access to cards; they may gain access to systems or provide cards and codes for analysis.

- *Insiders,* typically employees of the card issuer or scheme operator, have opportunities to copy, analyze, or steal data and hardware or to give special privileges or benefits to their friends *(sweethearting).*

- *Outsiders with system access,* for example maintenance engineers or telephone company employees, are often a particular threat as their activities are specialized and not fully understood by those supervising them.

- *Card thieves:* What can a person do with a card but no other information? Or what if they also know the identity of the cardholder and possibly some additional information held in a diary or wallet?

- *Criminal gangs* may gain access to dozens or hundreds of cards in a systematic way; they will also have access to computer systems and have the time to set up bogus accounts or businesses.

- *A serious and sustained effort to break the system for its own sake* is the most dangerous of all, but is only likely to be suffered by government or national financial systems. Political and religious fanatics are very dangerous adversaries, because they do not stop at the boundaries set by effort and reward.

As well as determining how easily each of these possible attackers could gain access to vulnerable elements of the system, the designer should consider how easily that penetration could be detected.

Quantitative analysis

Where we are able to set quantitative criteria, in the form of MTBIs or a Probability-Value for each type of incident, the design can now be tested against these criteria. This can be done most conveniently by setting them out against each other in a spreadsheet: For each risk we consider the possible causes and modes of failure, their likelihood, and the cost of the next level of countermeasure.

Armed with this information, we are able to state confidently not that the system is 100% secure but that we have calculated the risks and assured ourselves that further countermeasures are not justified by the risk.

Risk analysis checklist

Table 18.2 gives a simple checklist that may be used to analyze risk in a smart-card system. It is not comprehensive, and will need to be adapted for each situation, but it will give guidance to a designer seeking to analyze a system and to locate areas of high risk.

For each of these cases, a range of possible outcomes should be considered, and each should be assessed for severity and likelihood.

Table 18.2   Risk Analysis Checklist

Implementation	Operation
Design process	Hardware failures
Equipment and card selection	Software failures
Software design and selection	Card failures
Fitness for purpose	Card issuing
Testing	Use (consider each instance)
Delivery timescales	Lost and stolen cards
Implementation timescales	Expired cards
Project management	Counterfeit cards
Personnel	Breach of confidentiality—card data
Training	Breach of confidentiality—computer data
	Cardholder authentication: false accept/reject
	Card authentication: false accept/reject
	Terminal/system masquerade
	Key compromise
	Effects on other systems/customers

19

The Way Forward

I N THIS CLOSING CHAPTER we consider the strategies that major players could adopt in the coming years to promote the use of smart cards to enhance the security of data and systems.

Manufacturers

Semiconductors and masks

Chip manufacturers are already several steps ahead of the market in the maximum level of technology they can deliver. Economics alone dictate which products are offered to the market and at what prices. Smart-card chips set additional demands on the designer in terms of size, power consumption, and security, so that innovations developed for smart card ICs often find their way into other designs at a later stage. Semiconductor manufacturers that specialize in smart-card chips spend a significant proportion of their research budgets in this area.

The barriers to entry for other semiconductor manufacturers are therefore high; this restricts the number of companies in the market and allows them to set prices in a way that amortizes their high development costs. Existing manufacturers will want to maintain their lead, and the market is probably now delivering sufficient profit to justify the current level of research expenditure, in aggregate if not for every manufacturer. Potential market entrants must find a product in which they have an advantage and that would be attractive to the smart-card industry.

Foremost among these products are gate-array chips; here a small manufacturer can often gain an advantage by the level of service and turnaround offered in comparison with that of a larger company. We may therefore expect an increase in the number of gate-array-based smart cards.

This will be of interest to electronics design companies that have experience in gate-array design and can work in close concert with software security designers. This is the route most likely to provide differentiated products in the market.

Existing semiconductor manufacturers, on the other hand, will probably seek to provide more standard mask functions, particularly for high-end security chips. This will in some cases put them into competition with their own customers, but it is undoubtedly true that a detailed knowledge of the chip hardware features allows a much tighter operating system and security design.

Experience from the computer world suggests that despite the considerable pressure for standardization—of interfaces in particular—major improvements in security are usually made when a manufacturer breaks ranks and introduces a proprietary feature. This is followed by a period of consolidation as the new feature becomes standard, until the next breakthrough occurs.

One recent such breakthrough has been the introduction of an ITSEC-approved chip design; this is likely to set a pattern for other manufacturers to follow.

Cards

Card manufacturers usually seek multiple sources for their products; the number of qualified chip manufacturers is currently small and this sets limits on multiple sourcing. It is therefore in the interests of card manufacturers to seek a higher level of standardization, in particular at the interface to the mask. This is the same as the classic open-systems debate, which has been going on in mainstream computing for many years.

In practice it is likely that standardization can be achieved for memory cards and for low- and medium-functionality microprocessor cards. The more specialized high-performance and high-security chips will remain differentiated both in functions and in interfaces.

The market for card products is expanding rapidly; this growth is not only in higher volumes sold to existing sectors, but also in new sectors and large numbers of smaller projects. Card manufacturers will need to develop expertise in new vertical markets or work more closely with experts in those markets.

To provide an integrated secure service to their customers, card manufacturers and personalization companies will have to define the security functions they carry out, demonstrate to customers that they are able to maintain separation of cards and data from each of their customers, and demonstrate that they can meet their different requirements for security without compromise. Most of these companies rely heavily on physical security and procedures today, and these may need to be supplemented by provably secure computer systems, certified by an external body.

System designers and managers

Those responsible for designing and managing smart-card systems should follow the principles set out in this book, looking at all aspects of security and seeking management commitment to areas that are outside their direct control. They must not only look at the technical features of the system, but also at the environment in which it will operate. It will often be necessary to review or seek proof of the security of an external system, such as a network or a card-personalization system.

Designers must ensure that they know the specific characteristics of the systems and card products with which they are dealing, as well as the security features they offer. The system's functions and procedures must be set down clearly, so that everyone concerned with the implementation and operation of the system can understand them.

An organization should seek to develop a T-shaped knowledge base, where many people have some understanding of the technology and its implications, while a few will develop a much deeper and more detailed knowledge. The fast pace of development requires that anyone active in this field read the literature on a regular basis, and those responsible for security in particular would be well advised to read the encryption, security, and hacking newsgroups that deal with this subject.

Scheme operators

It takes much longer to build an infrastructure for smart cards than to develop individual applications: perhaps two to three years for a national infrastructure or five to ten years for a global one. Applications may be developed in six months and superseded in three years. Getting the infrastructure right, and making it easy to upgrade and add applications, should therefore be a top priority for any scheme.

Scheme operators in each sector need to agree on registration of schemes, issuers and applications, certification hierarchies, and procedures. Once one or two sectors have set up their schemes, this will set a pattern for others; to try to set up a full intersector scheme from the start would be too unwieldy.

Governments and intergovernmental bodies can help in this area, but ultimately it is the responsibility of the interested parties to define their requirements and to seek out the organizations best able to meet them. There is a pressing need for qualified institutions to act as key management centers and certification authorities.

A similar argument applies to hardware certification for open schemes. Here the need is for a common minimum specification to be applied to all hardware, which will ensure that cards cannot be damaged. An independent testing and certification body, or a hierarchy of test centers, will again be required.

Beyond smart cards

Smart cards are only a part of a continuous development towards more distributed computer systems that place the intelligence where it is required. Data can be stored anywhere in the world and accessed by communications links. The role of the card in this structure is as a proxy for the cardholder.

A smart card is a good ergonomic size: easily handled, not easily mislaid, and having a reasonable amount of space for visual features. But the most ergonomic feature of all is something we all carry all the time: ourselves.

When a single biometric can be shown to give a quick, noninvasive and reliable identification of a person's identity—probably through a DNA or other microbiological test—then the need for smart cards as an intermediary disappears. It is not inconceivable that the technology for this could exist within ten years.

The commercial, psychological, and infrastructural barriers will be much higher. It has taken 25 years for smart cards to reach the status of a widely accepted technology; it would take many times that to reach an international agreement on the use of a biometric as the sole and unchallengeable proof of a person's identity. Thus, smart cards have plenty of life in them yet.

Conclusions

As with governments, presidents, and marriages, technologies often go through a sequence of wariness and courtship, honeymoon and disillusionment before we finally learn to live with them and to benefit from them despite their limitations.

Managing expectations is the key to this process: public awareness of smart cards has been affected by a large amount of news coverage; much of this is exaggerated, sometimes to the point of fantasy. Claims that smart cards are "absolutely secure" were always bound to lead to disappointment. Smart cards are small and flimsy; systems are set up and managed by people, and people are fallible. The more we protect our data using invisible, electronic means, the more important the physical controls, procedures, and checks become.

None of this should detract from the knowledge that smart cards are, for many applications, the most economical and convenient way to store data and keys with an adequate level of security. As the technology develops, multiapplication cards and freely downloadable functions will become available, posing as many new problems as they solve.

Smart cards are here to stay, and will be used in increasing numbers of applications to provide security. Just as the marketing teams who think up smart-card schemes are able to free their imaginations, those responsible for designing systems must step beyond the details of the technology and match the product with the environment.

Appendix: Standards

THIS APPENDIX lists the most important international standards applicable to smart-card systems and their security.

Standard Name	Title and Description
ANSI X3.92	Data encryption algorithm (DEA). The main source for the DES algorithm.
ANSI X9.19	Message authentication check. The most widespread form of symmetric-encryption MAC.
ANSI X9.30-2 (1993)	Public-key cryptography using irreversible algorithms for the financial services industry. Part 2: Secure hash algorithm (SHA-1)
ANSI X9.31-1	Public key cryptography using reversible algorithms for the financial services industry. Part 1: The RSA signature algorithm. RSA is by some margin the most widely used public key encryption algorithm. Although the algorithm is patented, RSA Data Security licenses it for general use, and this standard gives recommendations in this respect.

CB	The French *GIE Cartes Bancaires* has defined a set of standards for use in France in connection with bank payment cards, all of which are chip-based. The scheme operates entirely within France.
	The CB standards include the Manuel de Paiement Electronique, which is a functional specification, and the CBSA, CBST, and PCEMA standards, which define the protocols. A further standard defines the mask (currently B0′) to be used in bank cards.
EN 726	Identification card systems—telecommunications integrated circuit(s) cards and terminals. This standard was produced by the ETSI working group TE9; its seven parts cover all aspects of the design of a multifunction IC card with the exception of the application itself.
EN1038	Identification card systems—telecommunications applications. Integrated circuit cards for pay phones. This standard, which was produced by CEN, complements EN 726.
EN 1546	Identification card systems—Intersector electronic purse. A development from the previous two standards but with banking industry and telecommunications industry involvement.
The EMV standards	Europay, Mastercard, and Visa integrated circuit card specifications for payment systems.
	The EMV standards define the content, structure, and programming of chip-based payment cards (including but not limited to credit, debit and, from issue 3, prepayment cards). They are designed to allow a single ATM or retail terminal to handle any bank-issued card without unduly restricting the ability of the card-issuing bank to define how its payment scheme should work.
	EMV Part 1: electromechanical characteristics, logical interface, and transmission protocols restates ISO 7816 part 3, restricting some of the options allowed by that standard.
	Part 2: data elements and commands corresponds roughly to ISO 7816 parts 4–6, defining the data elements and directories that applications may use.
	Part 3: transaction processing defines the procedures that must be implemented within the card and the terminal to carry out a payment system transaction in an international environment.
	The ICC terminal specification defines the mandatory, recommended, and optional requirements for terminals to support payment cards meeting parts 1–3. It is, however, a very low-level specification, and many features are at the discretion of the bank or terminal owner.
GSM	Global system for mobile communications. These standards are only available to members of the GSM consortium.

iKP	Internet keyed payment protocols. A set of protocols developed by IBM for secure payment, primarily using credit cards, across the Internet, using RSA public-key cryptography. iKP is the predecessor of SET (q.v.).
ISO 7498-2	Open systems interconnection (OSI) reference model: security architecture. The OSI model is the framework for most modern data communications standards. Although some of the most widely used standards do not comply with OSI recommendations in all details—the TCP/IP protocols commonly used in wide-area networking are not an OSI standard—the concept of a layered implementation, with specific services carried out at each layer, is still followed. The security architecture makes recommendations for the security services (from user authentication to nonrepudiation of delivery) that should be provided at each layer.
ISO 7810 (1985)	Identification cards—physical characteristics. This standard describes the shape, size, and environmental requirements of a plastic card to be used as an identification card. In practice, this standard size of card (85 × 54 mm) is used in almost all card applications.
ISO 7811	Identification cards—recording technique. The six parts of this standard cover embossing and magnetic-stripe recording.
ISO 7816-1 (1987)	Identification cards—integrated circuit cards with contacts. Part 1: physical characteristics. This covers all ISO 7810-sized chip cards with contacts, including memory cards and microprocessor cards. It includes specifications of the environmental and physical strength characteristics required.
ISO 7816-2 (1987)	Identification cards—integrated circuit cards with contacts. Part 2: dimensions and location of contacts. This standard defines the internationally accepted position for the chip (center left), but it also refers to the older "transitional" position still used in French bank cards today. The standard defines eight contacts, although many manufacturers only install six and many applications only use four or five.
ISO 7816-3 (1989) Amended 1992, 1994	Identification cards: integrated circuit cards with contacts. Part 3: electronic signals and protocols. This key standard defines the way a smart card communicates with the outside world. It includes the synchronous protocols normally used by memory cards as well as the asynchronous protocols more often used by microprocessor cards. ISO 7816-3 defines the structure of the *answer to reset* with which a smart card powers up and initiates communication with the terminal. It also defines the way a card should respond to over- or undervoltage. A further amendment to this standard is currently proposed to handle multiple input voltages (5V, 3V, and below

ISO 7816-4 (1995)	Identification cards—integrated circuit cards with contacts. Part 4: interindustry commands for interchange. This standard defines the content of several message types (including answer to reset). It also defines a file structure for use within a smart card, consisting of a master file (MF), dedicated files (DFs), and elementary files (EFs). The file and data object addressing provides a security architecture that is described in detail. It includes provision for applications to restrict access to whole EFs or data objects, based on either authentication or encryption.
ISO 7816-5 (1994)	Identification cards—integrated circuit cards with contacts. Part 5: Numbering system and registration procedure for application identifiers. This standard appoints KTAS in Denmark as the international body able to designate national authorities, which in turn can issue and authenticate application IDs.
ISO 7816-6 (1996)	Identification cards—integrated circuit cards with contacts. Part 6: Interindustry data elements. This is a directory of data elements for use in ISO 7816-4 applications.
ISO 7816	Part 7: Inter-industry commands for structured card query language (SCQL).
	Part 8: Security commands.
	Part 9: Enhanced commands.
	Part 10: Synchronous cards.
	Part 11: Security architecture, access control security attributes.
	These further parts to ISO 7816 are still being worked on in committee.
ISO 8583 (1993)	Financial transaction card originated messages—interchange message specifications.
ISO 8731	Banking—approved algorithms for message authentication. Part 1: data encryption algorithm; part 2: message authenticator algorithm.
ISO 9000 - 9004	Quality management and quality assurance standards.
	ISO 9001: Design, development, production, installation, and servicing.
	ISO 9002: Production, installation, and servicing.
	SO 9003: Final inspection and test.
	ㅐanking—PIN management and security.
	he two parts of this standard define the techniques and approved ｜gorithms recommended for handling PINs securely in a retail- ｜nking environment.
	｜curity techniques—digital signature scheme providing message ｜covery.

ISO 9797	Data cryptographic techniques—data integrity mechanism using a cryptographic check functions employing a block cipher algorithm.
ISO 9798	Security techniques—entity authentication mechanisms.
ISO 10118	Information security techniques: Hash Functions.
	Part 1: General principles.
	Part 2: Hash functions using an n-bit block cipher algorithm.
ISO 10126-1	Banking: procedures for message encipherment (wholesale).
	Part 1: General principles.
	Part 2: Data encryption algorithm (see also ANSI X3.92).
ISO 10202	Financial transaction cards—security architecture of financial transaction systems using integrated circuit cards.
	Part 1: Card life cycle.
	Part 2: Transaction process.
	Part 3: Cryptographic key relationships (draft).
	Part 4: Secure application modules.
	Part 5: Use of algorithms (draft).
	Part 6: Cardholder verification. Part 6 is mainly concerned with PIN verification, although an annex allows for passwords or biometric methods.
	Part 7: Key management (draft).
	Part 8: General principles and overview (draft).
ISO 10536	Identification card—contactless integrated circuit cards.
	Part 1: Physical characteristics.
	Part 2: Dimensions and location of coupling areas.
	Part 3: Electronic signals and reset procedures.
	Part 4: Answer to reset and transmission protocols.
	This standard defines a *close-coupled* card (up to 10 cm). A further standard (will be ISO 14443) for remotely coupled cards is in the early stages of drafting.
ISO 11568 (1994)	Banking—key management (retail). The three parts of this standard define the principles, techniques, and life cycle of keys in a symmetric cipher scheme. Further parts are proposed to deal with key management in a public-key environment.
PS/SC	The PC Smart Card architecture is an open architecture developed by a group of smart card and PC operating system vendors, notably CP8 Transac, HP, Microsoft, Schlumberger, and Siemens Nixdorf. It is intended to ensure interoperability between components from different vendors and across different hardware and software platforms. It comprises:

Part 1: Introduction and Architecture Overview.

Part 2: Interface Requirements for Compatible IC Cards and Readers.

Part 3: Requirements for PC-Connected Interface Devices.

Part 4: IFD Design Considerations and Reference Design information.

Part 5: ICC Resource Manager Definition.

Part 6: ICC Service Provider Interface Definition.

Part 7: Application Domain/Developer Design Considerations.

Part 8: Recommendations for ICC Security and Privacy Devices.

All of these documents can be downloaded from http://www.smartcardsys.com

SET
Secure electronic transactions: a set of standards for credit-card payment across networks, using the conventional cardholder–merchant–acquirer structure and the card number only. SET is a license-free open standard and can be downloaded from the Internet; acquirers, merchants, and cardholders obtain their keys from and are authenticated by a certification server operated by the card scheme. SET was developed jointly by CyberCash, GTE, IBM, MasterCard, Microsoft, Netscape, and Visa.

SEMPER
Secure electronic marketplace for Europe. SEMPER is a European research and development project in the area of secure electronic commerce over open networks, especially the Internet. At the time of writing, SEMPER has not published any standards, but it has close links with the SET standard and has published several working documents that show how SET may be used in an all-embracing payments system.

VIS
Visa integrated circuit card specification. This is a development from the EMV standards and gives more details of the schemes that will be supported by Visa.

Glossary

Application The program within a smart card that governs its external functions.

ATM Automated teller machine (cash machine).

ATM Asynchronous transfer mode (for data networks).

ATR Answer to reset: the data sent by a card to the reader when it is first powered up.

Authentication The process of verifying the identity and legitimacy of a person, object, or system.

CA Certification authority: a body able to certify the identity of one or more parties to an exchange or transaction.

CAM Card authentication method: the method (usually static or dynamic data authentication) used to verify that a card has come from a valid issuer and has not been tampered with.

Cardholder The person to whom a personal card was issued (not necessarily the person holding the card).

CEN *Centre Européen pour la Normalisation* (European Standards Center).

CLEF Commercial licensed evaluation facility: a body licensed to carry out security evaluations using the ITSEC criteria.

CRC Cyclic redundancy check: a check field often added to the end of a message, calculated as a polynomial from the rest of the message content. If a bit in the message is altered, then the CRC should alter.

CVM Cardholder verification method: the signature, password, PIN, or biometric used to check the identity of the cardholder, particularly for bank cards.

DDA Dynamic data authentication: authentication of a card using a challenge and response mechanism.

Digital signature An encrypted field, normally encrypted using the sender's private key, that is attached to a message to prove its source and integrity.

DSP Digital signal processor: an integrated circuit or specialized computer for processing high-frequency analog signals.

EEPROM Electrically erasable programmable read-only memory: semiconductor memory that retains its memory without power, but can be changed at any time.

EFT-POS Electronic funds transfer at point of sale: electronic payment.

Encryption   Manipulating data to make it unreadable to anyone who does not possess the decryption key.

EPOS   Electronic point of sale (terminal): a networked and programmable electronic cash register.

E^{2}PROM   See EEPROM.

ESD   Electrostatic discharge.

ETSI   European Telecommunications Standards Institute.

F/2F   Two-frequency encoding for magnetic stripes.

FAR   False accept rate: the percentage of impostors accepted by a biometric or other identification check.

Firewall   A device or system that intercepts traffic (for example, between the Internet and a private system) and filters out unwanted data or messages.

FPGA   Field programmable gate array: a semiconductor device that generates its outputs directly from its input states according to a "program" defined by the user.

FRR   False reject rate: the percentage of valid users rejected by a biometric or other identification check.

GSM   Global system for mobile communication: international standard for digital mobile telephony.

HSM   Host security module (or hardware security module): a hardware device used for storing keys and performing cryptographic functions under control of a host computer.

IC   Integrated circuit.

Data or message that has not been altered since it was

ITSEC   Information technology security evaluation criteria: European standard for evaluating the security of commercial computer products.

ITU   International Telecommunications Union: the international body responsible for telecommunications coordination, the successor body to CCITT. See also ETSI.

LCD   Liquid crystal display.

MAC   Message authentication check: a cryptographically derived block of data appended to a message to demonstrate that it has not been altered during transmission.

Mask   The fixed program of a smart-card microprocessor.

MTBF   Mean time between failures: a measure of the frequency of failures, but usually referring to a total failure.

MTBI   Mean time between incidents: a measure of the frequency of incidents, the inverse of the probability.

PGA   Programmable gate array (see also FPGA).

PIN   Personal identification number: a code (usually four to six digits) used as a password by a cardholder.

Public key   A public-key algorithm is one in which one key is published and the other kept secret.

PUK   PIN unblocking key (or personal unblocking key): a numeric code used to release a blocked application or card.

QA   Quality assurance.

RAM   Random access memory.

RFID   Radio frequency identification.

RISC   Reduced instruction set computer: a computer or microprocessor that, by operating with a smaller range of instructions, is able to achieve higher instruction speeds than conventional processors.

ROM   Read-only memory.

SAM   Security application module.

SDA   Static data authentication: authentication of a card by means of a signed copy of selected card data.

SIM   Subscriber identity module: the personalization chip card in a GSM telephone.

TCP/IP   Transmission control protocol/Internet protocol: the routing and addressing protocols respectively developed for the Internet and now used in many local- and wide-area networks.

TCSEC   Trusted computer security evaluation criteria: the U.S. *Orange Book* requirements for evaluating the security of computer systems.

Trojan horse   A program that runs without the knowledge of the system owner. It may collect or transmit data, make changes to the system, or simply make its presence felt in "benign" ways.

TTP   Trusted third party.

UPS   Uninterruptible power supply: a battery backup for use in case of mains failure.

Virus   A program (usually a trojan horse) that can replicate itself and thereby infect several systems from one source.

WORM   Write once read many times (form of semiconductor memory).

Bibliography

Smart card security references

The widest range of references on cryptography generally is to be found on the various Internet newsgroups that deal with this subject. They include *sci.crypt, comp.security,* and many subsidiary newsgroups under these. *Sci.crypt* includes a frequently asked questions (FAQ) file that can be downloaded, which includes a long list of references including surveys, history, and analytical work. *Alt.security* and *alt.hacker* frequently contain references to successful and attempted smart-card attacks.

Other references of specific interest to smart card security include:

Anderson, R. J., "Why cryptosystems fail," ACM Comms, Vol. 37, No. 11, pp. 32–40.

Anderson, R. J., and S. J. Bezuidenhout, "Cryptographic credit control in pre-payment metering systems," *IEEE Symposium on Security & Privacy,* 1995, pp. 15–23.

Anderson, R. J., and M. Kuhn, "Tamper Resistance—a Cautionary Note," *Proceedings of the 2nd Workshop on Electronic Commerce,* Usenix Association, California 1996.

Beller, M. J., and Y. Yacobi, "Fully-fledged two-way public key authentication and key agreement for low-cost terminals," *Electronics Letters (U.K.),* Vol. 29, No. 11, pp. 999–1001.

Burmester, M., Y. Desmedt, and T. Beth, "Efficient zero-knowledge identification schemes for smart cards," *Computer Journal,* Vol. 35, No. 1, pp. 21–29.

Frank, J. N., "Smart Cards Meet Biometrics," *Card Technology,* Sept/Oct 1996, pp. 30–38.

Fujioka, A., T. Okamoto, and S. Miyaguchi, "ESIGN: an efficient digital signature implementation for smart cards," *EUROCRYPT 91 Conf Proc.,* Berlin: Springer-Verlag, 1991.

Glass, A. S., "Why should secure cards be smart?" *Smart Card 2000 Conference Proceedings,* North Holland, Amsterdam, 1991, pp. 39–50.

Hendry, M., *Practical Computer Network Security,* Norwood, MA: Artech House 1995.

Hendry, M., "Security is More Than a Card Game," *Smart Card '97 Conference Proceedings,* QMS, Peterborough, England, 1997.

Holloway, C., "The IBM Personal Security Card," *Smart Card '91 Conference Proceedings,* Agestream, Peterborough, England, 1991.

Janson, P., and M. Waidner, "Electronic Payment Systems," *SI Informatik/Informatique (CH),* March 1995, pp. 10–15.

Konigs, H.-P., "Cryptographic identification methods for smart cards in the process of standardization," *IEEE Communications Magazine,* Vol. 29, No. 6, pp. 42–48.

Lim, C. H., Y. H.Dan, K. T. Lau, and K. Y.Choo, "Smart card reader," *IEEE Trans. on Consumer Electronics,* Vol. 39, No. 1, pp. 6–12.

Mondex International, *Mondex: Security by Design,* corporate publication, London, 1995.

Rhee, M. Y., Cryptography and Secure Communications, Singapore: McGraw-Hill, 1994.

Schaumüller-Bichl, I., "Card security: an overview," *Smart Card 2000 Conference Proceedings,* North Holland, Amsterdam, 1991, pp 19–27.

Schneier, B., "Applied cryptography: protocols, algorithms and source code in C," New York: Wiley, 1995.

Sociedade Interbancaria de Servicos, SA, *Multibanco Electronic Purse—Description of Scheme,* corporate publication, Lisbon, 1993.

Thomasson, J.-P., "Advances in Smartcard IC Technology," *SGS-Thomson Technical Article TA164,* 1996.

Waidner, M., "Secure Billing and Payment Over the Internet," *Rapperswil Networking Forum (CH),* 1995.

Wojciechowski, R., "Smart card issuing process," *Smart Card '91 Conference Proceedings,* Agestream, Peterborough, England, 1991.

About the Author

MIKE HENDRY has a degree in engineering from Cambridge and in business from IMI Geneva. He is multilingual and has worked in almost every country in Europe. For ten years he worked as a systems engineer, project manager, and consultant in real-time computing applications. Since 1982, he has been an independent consultant in data communications and payment systems, concentrating on marketing and business development for products and services on a European scale.

He works with banks, retailers, and service providers on a number of technology and business-strategy issues, including customer loyalty, card-fraud prevention, electronic data interchange, and the introduction of smart cards. He is currently involved in several smart-card and electronic-purse projects and in the development of secure satellite communications for transactional and Internet use.

Index

The Artech House Telecommunications Library

Vinton G. Cerf, Series Editor

UNIX Internetworking, Second Edition, Uday O. Pabrai

Videoconferencing and Videotelephony: Technology and Standards, Richard Schaphorst

Wireless Access and the Local Telephone Network, George Calhoun

Wireless Communications in Developing Countries: Cellular and Satellite Systems,
    Rachael E. Schwartz

Wireless Communications for Intelligent Transportation Systems, Scott D. Elliot and
    Daniel J. Dailey

Wireless Data Networking, Nathan J. Muller

Wireless LAN Systems, A. Santamaría and F. J. López-Hernández

Wireless: The Revolution in Personal Telecommunications, Ira Brodsky

Writing Disaster Recovery Plans for Telecommunications Networks and LANs,
    Leo A. Wrobel

X Window System User's Guide, Uday O. Pabrai

For further information on these and other Artech House titles, contact:

Artech House
685 Canton Street
Norwood, MA 02062
617-769-9750*
Fax: 617-769-6334*
Telex: 951-659
email: artech@artech-house.com

Artech House
Portland House, Stag Place
London SW1E 5XA England
+44 (0) 171-973-8077
Fax: +44 (0) 171-630-0166
Telex: 951-659
email: artech-uk@artech-house.com

WWW: http://www.artech-house.com

* *As of September 1, 1997, new area code is (781)*